A Guide to Laboratory Design

K. EVERETT BSc, ARIC,
University Safety Officer and Honorary
Lecturer in the Department of
Community Medicine and General
Practice in the University of Leeds

D. HUGHES BSc, PhD, CEng, FInstP, MIEE,
University Radiation Protection Officer and
Associate Lecturer in Physics in the
University of Leeds

Butterworths   London & Boston

THE BUTTERWORTH GROUP

ENGLAND

Butterworth & Co (Publishers) Ltd
London: 88 Kingsway, WC2B 6AB

AUSTRALIA

Butterworths Pty Ltd
Sydney: 586 Pacific Highway, NSW 2067
Melbourne: 343 Little Collins Street, 3000
Brisbane: 240 Queen Street, 4000

CANADA

Butterworth & Co (Canada) Ltd
Toronto: 2265 Midland Avenue,
   Scarborough, Ontario, M1P 4S1

NEW ZEALAND

Butterworths of New Zealand Ltd
Wellington: 26–28 Waring Taylor Street, 1

SOUTH AFRICA

Butterworth & Co (South Africa) (Pty) Ltd
Durban: 152–154 Gale Street

USA

Butterworth
161 Ash Street, Reading, Boston, Mass. 01867

All rights reserved. No part of this publication may be reproduced or transmitted in any form or by any means, including photocopying and recording, without the written permission of the copyright holder, application for which should be addressed to the publisher. Such written permission must also be obtained before any part of this publication is stored in a retrieval system of any nature.

First published 1975

© K. Everett & D. Hughes, 1975

ISBN 0 408 70682 1

Typeset in England by Cold Composition Ltd, Tunbridge Wells, Kent

Printed in England by R. J. Acford Ltd, Chichester

Preface

The authors wonder what reviewers will make of this book; it is not on architecture, or on engineering or on science. Its purpose is to draw together information which was in the main sought of the authors by many people involved in various aspects of laboratory design. It is presented in the hope that it will be of help to others who are engaged in the designing and the building of laboratories.

K.E.
D.H.

To the reader

This book attempts to set out many of the problems encountered in the design of laboratories for work with hazardous materials and to describe some of the solutions possible, often from the direct experience of the authors. However, the authors are not omniscient; the advice is given in good faith, but other solutions may be preferable in some instances. *In all cases, a designer must satisfy himself that any arrangement that he proposes to adopt is suitable and safe for his specific application.*

Acknowledgements

Laboratory design is an exercise which involves many people from several disciplines. The authors record their appreciation of the many helpful discussions which they have had with various colleagues during the past few years. They are particularly indebted in this respect to the staffs of the Planning Office and of the Office of the Surveyor of the Fabric in the University of Leeds.

The authors wish to thank the following persons, organisations and journals for their permission to reproduce drawings and photographs for which they hold the copyright:

Chemistry in Britain
Dunford Fire Protection Services Ltd
Engineering Developments (Farnborough) Ltd
Professor S. C. Frazer, University of Aberdeen
Mr Denys Horner, Senior Assistant Bursar (Planning), the University of Leeds
Koch-Light Laboratories Ltd
Laboratory Animals Ltd
Laboratory Practice
Microflow Ltd
The Fire Protection Association

Finally, the authors wish to acknowledge their appreciation of Mrs J. M. Carr and Miss D. Lewis, who translated the manuscript into a legible typescript for transmission to the Publishers.

Contents

Chapter One: Introduction   1

Chapter Two: Laboratory Suites   4

Chapter Three: Basic Design Features   10
   3.1 Walls and ceilings   10
   3.2 Floors   12
   3.3 Working surfaces   16
   3.4 Laboratory furniture and fittings   20
   3.5 Services   26

Chapter Four: Fire Precautions   31
   4.1 Consultation   31
   4.2 General principles   32
   4.3 Building layout and emergency escape routes   34
   4.4 Fire alarms and fire detectors   36
   4.5 Computers   38
   4.6 Problems associated with flammable and toxic solvents and with explosives   38

Chapter Five: Means of Detecting and Extinguishing Fires   40
   5.1 Automatic fire detectors   40
   5.2 Installation of detectors   43
   5.3 Fire extinguishers   44
   5.4 Classification of fires, methods of extinguishing and choice of extinguisher   46

Chapter Six: Laboratory Ventilation   50
   6.1 Methods of providing ventilation   50
   6.2 General dilution ventilation   50

6.3	Local exhaust or spot ventilation	53
6.4	Partial enclosures (fume-cupboards)	55
6.5	Special enclosures	67
6.6	Total enclosures (glove-boxes)	68
6.7	Air inlet systems	72

Chapter Seven: Fume Extraction and Dispersal — 76

7.1	The extract system	76
7.2	Filtration	78
7.3	Air-flow sensors	79
7.4	Fire-dampers	80
7.5	The ductwork	82
7.6	The extract fan	85
7.7	Fume dispersal	87

Chapter Eight: Laminar Air-flow Clean Rooms and Work Stations — 93

8.1	The need for laminar air-flow	93
8.2	Basic design features of laminar air-flow rooms	95
8.3	Specifications for clean rooms	96
8.4	Types of laminar air-flow clean room and work station	97

Chapter Nine: Stores and Other Ancillary Areas — 104

9.1	Supporting areas	104
9.2	Solvent stores	105
9.3	Solvent dispensaries	110
9.4	Waste solvent facilities	111
9.5	Gas cylinder stores	111
9.6	Chemical stores	114
9.7	Radioactive stores	115
9.8	Strong-rooms	117
9.9	Waste disposal	117
9.10	Stores compounds	119

Appendix A: The Requirements for Work with Radioactive Substances — 121

A.1	The nature of the hazard	121
A.2	Legislation and codes of practice	124
A.3	The grading of radioactive laboratories	126

Appendix B: The Requirements for Work with
              Microbiological Materials   128
  B.1 The nature of the hazard   128
  B.2 Methods of sterilisation   129
  B.3 Microbiological safety cabinets   130
  B.4 Waste disposal   133

Appendix C: Threshold Limit Values for Chemically
              Toxic Materials   135

Appendix D: Carcinogenic Substances   137

Appendix E: School Laboratories   138

Appendix F: Fire Offices' Committee Rules   139

References   140

Index   153

One
Introduction

The dangers associated with the use of hazardous substances in research and in routine laboratory work are greatly reduced when the operations are carried out in laboratories properly designed or adequately adapted for such work. Hazardous substances may be grouped under a number of headings, not necessarily mutually exclusive. They may be: allergenic, asphyxiant, carcinogenic, corrosive, dermatitic, explosive, flammable, lachrymatory, pathogenic, poisonous, powerfully oxidant, powerfully reducing, radioactive, teratogenic. In most cases laboratory design must take account of the same basic features: containment, cleanliness, ventilation, waste disposal, storage, security, control of personnel, fire precautions and provisions for an emergency. In certain cases additional special features are needed: for example, shielding for radioactive substances or sterilisation for microbiological materials.

In this book, in order to avoid a great deal of repetition, the basic features of laboratory design are considered first (Chapters 2, 3, 4 and 5). The various aspects of the control of airborne contaminants by ventilation are dealt with next (Chapters 6, 7 and 8). Stores and other ancillary areas are considered in Chapter 9. Finally, the additional special features required for some of the different groups of hazardous materials are summarised in the Appendices.

No attempt is made to provide an architecturally integrated design for a whole laboratory and office complex, and the reader is referred to other works for this aspect of the problem.[1-9] In most chapters the authors have tended to concentrate on points of detail rather than present over-all schemes. The extent to which these points need to be

Introduction

incorporated in the design of a particular laboratory and office complex depends on the nature of the work to be done in the various sections of the complex. Much chemical work, for example, is relatively innocuous and, although some of the design features described are appropriate in a laboratory for such work, it is not necessary to provide others, such as rigid segregation of personnel and glove-box facilities. On the other hand, there are circumstances when, for example, the work involves virulent pathogenic micro-organisms or alpha-particle-emitting radionuclides, where a segregated suite of rooms with elaborate mechanical services is essential for the safety of people and for the successful control of the experimental procedures. With these circumstances in mind, Chapter 2 has been devoted to the design of a *laboratory suite* where hazardous experiments can be contained; such a suite might be provided as one special area within a large building with many other laboratories for less hazardous work, or the suite might be expanded to become a whole laboratory block if the scale of the work required this.

It must be emphasised that detailed consideration by the architect, the engineer and the client of many of the matters raised in this book is imperative right from the conception of the building project to the final commissioning. Laboratories for work with hazardous materials are usually complicated entities with extensive engineering services; if the laboratories are to function efficiently and safely, the services must be considered from the outset of the project and the appropriate provisions must be made in the over-all architecturally inspired design to accommodate them so that their efficiency is not impaired. The classic example here is that of the fume-cupboard and its extract system; if the decisions relating to the location of a fume-cupboard and to the routing of its extract ductwork are left to a late stage in a project, the most hideous and undesirable convolutions can occur in the ductwork, as installed, with consequent difficulties if the extract fan has been sized on the basis of a lower duct-resistance.

No mention is made of cost/effectiveness analysis for any of the provisions suggested. What is very clear, however, is that it is cheaper to build the laboratory correctly at the first attempt than to have to rectify deficiencies after the laboratory has been brought into use. Equally clear is the fact that

Introduction

architects and engineers usually have to work within prescribed cost limits which may restrict the extent of the facilities that can be provided. On the latter point, if the funds available are inadequate to provide the facilities that are necessary for the work to be carried out, then the project should be reconsidered.

Certain aspects of laboratory design are subject to statute in the UK and to similar legal constraints in other countries; for example, there are regulations relating to fire precautions and to radioactive substances. Further control over design is exercised by national codes of practice and by the recommendations of international bodies; these do not have the force of statute law, but usually they can be invoked in civil actions at law. Where appropriate, statutory requirements and recommendations applicable in the UK are usually referred to in later chapters. It is possible that the present interest in health and safety in the workplace shown by the UK government and by other bodies will result in further control of the standards required in the design of laboratories; of particular interest will be the consequences of the Health and Safety at Work etc. Act.[10]

Two
Laboratory Suites

It is sometimes appropriate, because of the level of hazard involved, to confine work with hazardous materials to a segregated suite of rooms in a building or to provide a separate laboratory block if the scale of the work requires it. Such arrangements have been made for many years for work with radionuclides[11-14, 216] and they are applicable to other types of work also.[6, 15]

When such a laboratory suite is planned, special consideration must be given to the following points, which are illustrated schematically in *Figure 2.1*:

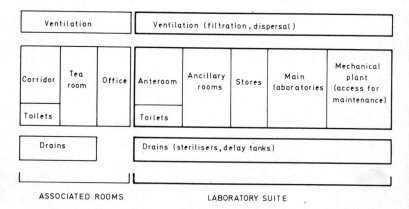

Figure 2.1 Schematic diagram of the general arrangement for a laboratory suite

1. The segregation of the laboratory suite from other (non-laboratory) areas.
2. The control and the suitability of the access to the suite.

Laboratory Suites

3. The separation of the various levels of potential hazard within the suite.
4. The provision of individual laboratories and ancillary rooms as appropriate for the operations to be carried out.
5. The provision of alternative exits for use in an emergency.
6. The design of the ventilation and drainage systems and the space necessary to accommodate them.
7. The need to install services so that much of the maintenance can be carried out without high-hazard areas having to be entered.
8. The provision of associated rooms outside the suite, such as tea rooms and offices.

An isolated single-storey building devoted entirely to the one type of work is an attractive proposition, provided that it will not be surrounded by taller buildings at a future date. Among the advantages are:[16]

1. The less serious consequences that are likely to arise if a temporary failure to contain the hazardous material occurs.
2. The easier control of persons entering the area.
3. The absence of other occupants who might be affected by the discharge of fumes.
4. The shorter runs of internal drains to a relatively large sewer with the consequent advantage of early dilution.
5. The higher permissible floor-loading available to support heavy items such as the shielding against radiation required in some radioactive laboratories.
6. The reduction of the problems caused by vibration.
7. The greater freedom possible in the arrangements for the delivery of materials and the removal of hazardous waste.

However, in many places it is inevitable that the work will have to be done in part of a tall multistorey block shared with other occupants. If the laboratory suite is placed near ground level, long, expensive extract ducts will be needed for the ventilation system. These will have to discharge through a

Laboratory Suites

high stack, possibly appearing as an incongruous feature disturbing a graceful skyline, in order to escape from the downdraughts surrounding the building (see Chapter 7). If the laboratory suite is sited on the top floor, the long extract ducts are exchanged for long drain runs with the attendant risk of contaminating uncontrolled areas lower down in the building; deliveries of materials, perhaps heavily shielded in the radioactive case, and the removal of contaminated waste are less conveniently arranged.

Figure 2.2 illustrates an arrangement of rooms which could be incorporated in a building where a small number of people wish to use a particular type of hazardous material — for example, radioactive substances. The various stages of the work are identified and appropriate provisions made for them: the stocks are kept in the stores; the initial dispensing at relatively high concentrations and the more hazardous manipulations are performed in the glove-box, safety cabinet or fume-cupboard; the less hazardous work is accommodated on the open benches; the samples of very low activity level are manipulated or assayed in a separate ancillary room; the staff washing facilities and the personal lockers for use by staff who wear protective clothing in the laboratories are accommodated in the changing room which is divided into 'clean' and 'dirty' sides by a change barrier; and alternative exits are provided from each room and from the whole suite.

There are incidental advantages in the type of arrangement shown in *Figure 2.2*. The changing room acts as a buffer between the main laboratories and the corridor and it discourages the casual caller (complete with tea mug, pipe or cigarette). The ancillary room can provide some protection for expensive electronic (or other) equipment which could be easily damaged by exposure to any corrosive fumes generated if a relatively small fire occurred in one of the main laboratories. The ancillary room can also be used for purposes of calculation and writing-up with the workers in the main laboratories kept under surveillance (in case of accident) through the observation window.

The laboratory suite illustrated in *Figure 2.2* represents the basic unit which is borne in mind throughout this book. Its size obviously varies with the circumstances, but ample space should always be provided to permit of safe working. A minimum area of 4.5 m^2 (50 ft^2) per person has been recom-

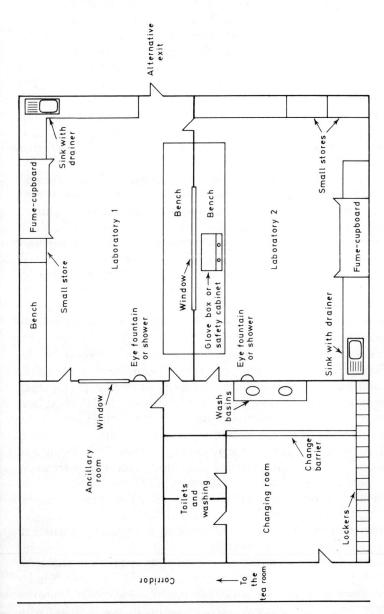

Figure 2.2 Illustrative arrangement of a laboratory suite

Laboratory Suites

mended in the actual laboratory,[16] and when glove-boxes or safety cabinets are used, sufficient clearance must be allowed to enable a worker to withdraw his hands from the gloves (Section 6.6). In planning the subdivision of a laboratory suite, the concept of the laboratory space unit can be used;[7] this is a module of laboratory space, either an open bay or an enclosed room, which accommodates a small group of workers or an arrangement of equipment. For a rectangular unit, a useful size for the laboratory space unit, based largely on the limits of human reach, is 6.1 m × 3.3 m, which gives an area of 20.5 m^2.

In some circumstances the provision of a whole suite may not be justified; but the underlying principles of segregation and containment should be observed. On the other hand, if the work is to be extensive, more fume-cupboards, glove-boxes or safety cabinets, additional laboratories or ancillary rooms, a separate store and a 'dirty' office may be required.

Adjustments will also be necessary which take account of the precise nature of the hazardous material to be used (Appendices A and B); for example, heavy shielding may be required in radioactive laboratories and sterilising facilities in microbiological laboratories. If the suite is to be in a hospital,[17] clinically oriented rooms may be necessary: for example, rooms for the preparation (sometimes under sterile conditions) and the administration of doses to patients; laundries for dealing with contaminated bedclothes; and stores for excreta awaiting disposal.[7,11,14,18,216] If animals are involved in the experiments, an arrangement such as that illustrated in *Figure 2.3* may be appropriate:[19,20] 'clean' and 'dirty' corridors are provided; dispensing and related chemical or similar operations are performed in the laboratory; various ancillary rooms open off either the 'clean' or 'dirty' corridor as appropriate; and the animal accommodation is so arranged that each experiment is segregated in a 'roomlet' in order to reduce the risk of cross-contamination and to limit the area involved and the number of persons at immediate risk if an accidental release of hazardous material occurs. Similar arrangements can be used for microbiological laboratories with additional features such as air-locks, pass-hatches with appropriate means of sterilisation and more double-ended autoclaves.[6,15,21,22]

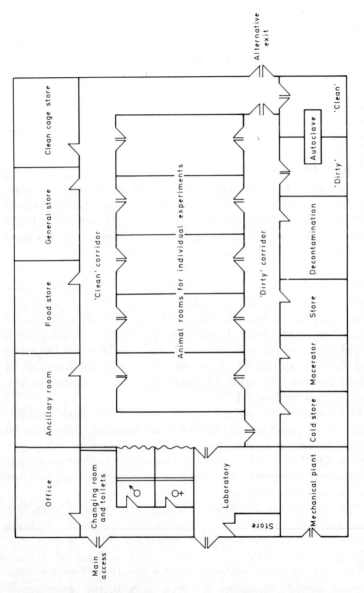

Figure 2.3 Illustrative arrangement of a suite for experiments involving animals.[19] *(Courtesy of Laboratory Animals Ltd)*

Three
Basic Design Features

3.1 Walls and ceilings

The principal features required of walls and ceilings are that: (a) they should possess a smooth, impervious, readily cleaned surface to minimise the accumulation of dust and to facilitate decontamination; (b) they should be vermin-proof; (c) they should have a slow flame-spread characteristic.

The first two features are easier to achieve if the walls and ceilings are finished with plaster. The plaster can then be coated with a paint which will withstand washing without staining or being otherwise adversely affected by the decontaminating or germicidal agents used. If walls are constructed of a dry form of partitioning, movement of the timbers can result in cracks and fissures appearing. The use of plastics in buildings is increasing but, because of their poor fire resistance, the applications are limited.[23]

Occasionally the paints applied to surfaces have to withstand more exacting conditions than washing; they may be required, for example, to protect combustible substrates against fire[24] or to have a low flame-spread[25] or to possess good resistance to specific chemicals or to be easily stripped for decontamination purposes.

The properties of paint coatings vary greatly with the chemical composition; in consequence, different products within the same general classification may show appreciable differences in chemical resistance and in ease of decontamination. General guidance on the selection of coatings for ease of decontamination is given in a British Standard.[26] Gloss finishes are preferred because they contain a relatively low proportion of pigment and present a surface which is smooth and free from pores.

Basic Design Features

For surfaces requiring only occasional decontamination with mild detergents, the conventional alkyd resin type of paint is suitable provided that it is not used under alkaline conditions. Three types of paint which are more resistant chemically are: (1) chlorinated rubber, (2) two-pack epoxy and (3) two-pack polyurethane.[26]

Chlorinated-rubber-based paints have been used extensively. The binding medium is a mixture of chlorinated rubber and plasticising resins. The plasticisers must be chemically inert in order to give a good chemical resistance. The resistance of these paints to most aqueous solutions is good, but they dissolve in or are softened by a number of organic liquids. Their maximum continuous working temperature is about 60°C.

Epoxy-resin-based paints have good chemical resistance to alkalis and to many organic liquids, but they blister in pure water[26] and are attacked by concentrated mineral acids. They have some abrasion resistance and their maximum continuous working temperature is about 90°C.

Polyurethane paints (Section 3.3) have a very wide range of compositions. Their properties therefore vary considerably, but, in general, may be taken to be similar to those of epoxy resin paints.

Occasionally walls and ceilings in radioactive laboratories need to be thickened to provide shielding against gamma-ray-emitting substances. This usually occurs only when there is a specific requirement at, say, certain locations in the main dispensing area, or behind fume-cupboards or around stores, and expert advice is necessary for such designs (see Appendix A). When walls are thickened it would be useful if some identifying feature could be incorporated which would make them readily distinguishable from ordinary walls, even after the passage of many years when all the original staff may have left.

Structural simplicity is desirable in order to prevent the accumulation of dust and to facilitate the decontamination of surfaces. Ledges should be avoided; inside rooms they harbour dust and outside buildings they act as roosting perches which are undesirable, particularly near animal or microbiological laboratories. Pipes, conduits, ducting, fuse-

Basic Design Features

boxes and other projecting items are better deployed outside the hazardous areas.

Special attention should be given to the manner in which the pipes and conduits pass through the wall, ceiling or floor; if the sealing is inadequate, dirt-laden air may enter, fumigation may be less effective and two-way traffic of undesirable livestock may thrive. It is advantageous to group services so that several enter the laboratory through one demountable wall or ceiling panel; at a later date, additional services can be introduced into the laboratory without the disturbances associated with the cutting of holes through brick and plaster.

Special attention should be given during the initial planning to any problems that may arise as a result of the mode of heating. In particular, any hot-water radiators used should be of patterns which are easily cleaned (e.g. flush-fitting or embedded panels or types with sufficient clearance to walls to allow of cleaning behind them) and recirculation of extracted air is not usually permissible (see Section 6.7).

3.2 Floors

Floors should not flex under load and therefore are constructed preferably with a concrete base. The floor covering must be carefully sealed so that spillages and flooding can be contained; the consequence of flooding in a multistorey building without sealed floors can be serious even if the water is uncontaminated. In some laboratories, where extensive washing with decontaminating solutions is carried out, a controlled floor drain with appropriate grading of the floor level may be required.

The floor covering should be jointless, or, if made of a material in sheet form, the joints should be carefully welded. Particular care with the sealing is necessary where any fixtures, such as the legs of benches, have to be secured to the floor through the covering; in some cases it may be sufficient to coat the securing screws with a sealant and in others it may be preferable to form a cove or sleeve around the fixture. The covering should be carefully bonded to the sub-floor over the entire area so that, if the covering is punctured, liquids cannot spread out extensively beneath it.

Table 3.1 SUMMARY OF RESISTANCE TO CHEMICAL ATTACK OF FLOOR-COVERINGS[13] (24 h EXPOSURE)

Attacking agent	Linoleum[27]	'Crestaline'[27] 67% PVC	Lefco fully vitreous ceramic tile[28]
Acetone	2	3	1
Animal fats	1	1	1
Beer*	1	1	1
Butyl alcohol	2	2	1
Carbon tetrachloride	2	3	1
Chloroform	2	3	1
Chromic acid 3N	2	1	1
Diethyl ether	2	2	1
Ethyl alcohol	1	1	1
Glacial acetic acid	2	1	1
Hydrochloric acid	2	1	1
Hydrogen peroxide (10%)	1	1	1
Mineral oils	1	1	1
Nitric acid 7N (36%)	2	1	1
Paraffin	2	1	1
Petrol	2	1	1
Potassium hydroxide	3	1	2
Sodium hydroxide	3	1	2
Sulphuric acid 10N (38%)	2	1	1
Trichloroethylene	2	3	1
Vegetable oils	1	1	1
Water	1	1	1
Xylene	2	3	1

*Drinking is not permitted in radionuclide laboratories.
1 = satisfactory; 2 = slight attack; 3 = attacked.

Table 3.2 EASE OF DECONTAMINATION OF THE SURFACE OF VARIOUS MATERIALS[29, 30]

Material	Ease*
Stainless steel	0.01
Industrial polyvinylchloride, grey	0.01
Industrial polyvinylchloride, white	0.05
Polypropylene on glass fibre base	0.2
Plastics laminate	1.5
Plastics laminate with abraded surface	1.4
Plastics laminate treated with hypochlorite abrasive cleaner	5.7
Plastics laminate, aged	4.5
Polyurethane-varnished hardwood	33.2
Vinyl flooring	0.03
Vinyl flooring plus asbestos filler	4.4
Vinyl flooring plus rubber filler	10.0
Linoleum	6.9

*Percentage of contaminant (^{60}Co/^{134}Cs) remaining after decontamination by British Standard procedure.[31]

Basic Design Features

The following are among the materials used for the floor covering in a laboratory; some guidance is available on their resistance to chemical attack (*Table 3.1*) and on their ease of decontamination (*Table 3.2*).

Vinyl sheet. Vinyl material in sheet form, but not as tiles because of the large number of joints that would be present, is often used. For satisfactory decontamination, the material should contain a high proportion of polyvinylchloride;[32] figures of at least 30% by weight[26] and 50%[12] have been recommended. The ease of decontamination also depends on whether a rubber or an asbestos filler is incorporated (*Table 3.2*). The material is sufficiently flexible for it to be formed into an integral cove at walls and other fixtures and any joints necessary can be welded by a hot-air process. It should not be laid where it would be subjected to abrasive materials or to heavy point loads and it will soften if exposed to organic liquids for long periods (*Table 3.1*).

Linoleum. Good-quality cork linoleum may be used. It is necessary to keep it well polished[26] to assist decontamination, which is not as easy to effect as in the case of a vinyl floor covering with a high content of polyvinylchloride (*Table 3.2*). A separate pre-formed coved skirting has to be fitted, because linoleum is not sufficiently flexible to be formed into an integral cove, and joints have to be cold-welded. Linoleum is not recommended for rooms which are to be washed frequently with disinfectant.[6] The resistance to chemical attack is summarised in *Table 3.1*.

Vitreous ceramic tiles. In some cases (e.g. in some animal rooms and in pilot plants) squared-off fully vitrified ceramic tiles may be used.[13, 26] It is important that the tiles be fully vitrified throughout the body of the tile; this ensures a low water-absorption, and chipping and surface defects do not significantly impair the decontamination properties, which are rated as good.[33] The tiles should be pointed to the tile depth in a suitable non-porous resin cement containing chemically inert filler, previously tested for its decontamination properties. The chemical resistance of ceramic tiles is summarised in *Table 3.1*.

Synthetic resin screeds. Screeds based on epoxide and polyester resins and using siliceous or aluminous fillers are often satisfactory; they are particularly useful where heavy

Basic Design Features

loads are to be used or where the substrate is uneven.[26] Chipping and surface defects do not significantly impair their decontamination properties and they can be formed into coves. In some cases it is possible for a similar resin-based paint to continue from the floor coving up the walls.

Figure 3.1 *Vinyl sheet floor covering with integral coving at the walls and with the joints between sheets welded*

Mastic asphalts. Asphalts are composed basically of two constituents — the rock aggregate and the bituminous binder.[32] These two constituents are present in a variety of natural deposits; in addition, crushed rock aggregates and the bituminous residues of the petroleum industry are used. From this range of starting materials various grades of flooring material are made. They are hard-wearing, dustless, non-slip and impervious to water; by using a siliceous aggregate and a binder selected for chemical resistance, an asphalt can be produced which is resistant to dilute acids and alkalis. These materials are usually laid at least 20 mm thick, can be formed into coves and will withstand some movement in the sub-floor. The main disadvantages are the ease of deformation under heavy loads, the susceptibility to temperature changes, the adverse effects of oils and solvents, and the difficulty of removing particulate contamination which has become embedded in the asphalt. The application of a suitable polish to the surface is beneficial.[26]

For ease of cleaning, coves or rounded corners should be formed at all the intersections between ceilings, walls and floors in laboratories. However, because of the cost involved, often only the coves between the walls and the floors are provided (*Figure 3.1*); these are usually the most important ones, because they are intended to contain liquid spillages and to make mopping of the floor easier. Generally, a cove about 75 mm high is sufficient and it can be continued across selected doorways as a *gently* rising hump, say 30 mm high, in a contrasting colour, provided that such a hump does not interfere with the movement of apparatus.

3.3 Working surfaces

The properties of an ideal working surface include: (a) a hard, scratch-resistant surface, (b) low porosity, (c) good heat resistance, (d) good chemical resistance, (e) good resistance to staining, (f) ease of decontamination, (g) availability as a virtually continuous surface or in large sheets, and (h) reasonable cost.

Traditionally, hardwoods have been used for bench surfaces in chemical laboratories, but they are not well suited for work with hazardous substances or when techniques require low levels of contamination. Changes in the moisture content of solid timber can cause sufficient movement to open up cracks several millimetres wide in the surface,[3] and hardwoods treated with polyurethane lacquer are very difficult to decontaminate (*Table 3.2*).

An early survey of surface materials for use in radiochemical laboratories[34] provided much data on products commercially available in the USA. Plastics have found many applications but are restricted to temperatures below 155°C (in some cases as low as 40°C) and some burn readily.[35]

In the UK, materials such as the following are generally used (*Tables 3.2* and *3.3*):

Basic Design Features

Melamine resin plastics laminate, such as Formica,[37] will withstand temperatures up to 154°C (310°F). It should be bonded to a waterproof backing material, such as resin-bonded plywood or particle board, with a resin glue. Contact adhesive may not withstand a sufficiently high temperature.

Polyvinylchloride sheet, such as Darvic,[38] has a service temperature range from −30°C (−22°F) to 60°C (140°F). The low figure for the maximum temperature precludes its use in some applications where hot vessels may be placed on the bench or where there is significant heat reflected onto the bench surface. However, this disadvantage can be overcome by supporting hot items on thick asbestolite blocks, raised 10 or 20 mm above the level of the bench, with the asbestolite protected from contamination by a replaceable covering of aluminium foil.[30] Darvic can be welded and is self-extinguishing if in a fire, but its decomposition products are corrosive (see Section 6.4.1).[84-89,215] It is fairly soft and may be scratched easily.

FMB grade stainless steel[39] is a suitable metal for a working surface where metals are acceptable. It is less susceptible to chemical attack than the '18/8' grade of stainless steel used for domestic purposes, but is prone to attack by hydrochloric acid and by certain other chemicals.

Glass-fibre-reinforced resin, such as bisphenol polyester resin,[36] can be moulded to shape and can be used at temperatures up to 95°C (203°F), although the softening point is about 135°C (307°F). The material can be treated to make it fire-retardant, but it may burn in a fire with the production of a considerable amount of corrosive smoke.

Polypropylene[40] can be welded and heat-formed, and softens at a temperature of 145°C (293°F). Because of the high resistance to chemical attack, there may be difficulty in finding a suitable adhesive for it, and the advice of the manufacturers should be sought. The material is not fire-resistant and, once ignited, it will continue to burn. Its surface is less susceptible to scratching than that of polyvinylchloride.[30]

Melamine and polyurethane lacquers[41] can be applied to a hardwood bench-top. This type of finish is advocated by some workers for work with low levels of potential contaminants;[33] however, polyurethane-varnished hardwood is reported to have very poor decontamination properties.[30]

Table 3.3 SUMMARY OF RESISTANCE TO CHEMICAL ATTACK OF MATERIALS FOR WORKING SURFACES[13,36]

Attacking agent	Formica melamine resin plastics[37]	PVC[38]
Acetic acid (glacial)	1	2
Acetic acid (60%)	1	1
Acetone	1	3
Ammonium hydroxide	1	1
Aniline	–	3
Aqua regia	1	–
Beer	–	1
Benzene	–	3
Benzylalcohol	–	3
Bromine	–	3
Carbon disulphide	1	2
Carbon tetrachloride	1	2
Chlorine (moist)	–	2
Chlorobenzene	1	3
Chloroform	1	3
Chromic acid (80%)	2	1
Chromic/sulphuric acid	–	–
Cresylic acid	1	–
Diethyl ether	1	3
Ethyl acetate	1	3
Ethyl alcohol (96%)	1	1
Formic acid	1	1
Hydrochloric acid (conc.)	1	1
Hydrofluoric acid (60%)	2	2
Hydrogen peroxide (30%)	–	1
Iodine in KI solution	1	3
Nitric acid (95%)	2	3
Nitric acid (50%)	2	1
Petroleum ether	–	–
Phosphoric acid (95%)	1	1
Potassium hydroxide	1	1
Sodium hydroxide	1	1
Sulphuric acid (98%)	2	2
Sulphuric acid (60%)	1	1
Toluene	–	3
Trichloroethylene	–	3
Xylene	–	3

Table 3.3 (*continued*)

Polypropylene [40]	FMB stainless steel [39]	Bisphenol polyester resin [36]	Acid-catalysed melamine lacquer [41]	Polyurethane lacquer [41]	Pyroceram [42]
1	1	1	2	2	1
1	1	1	1	1	1
1	1	2	2	2	1
1	1	1	1	1	1
1	–	3	2	1	1
2	–	–	1	1	–
1	1	1?	–	–	–
2	–	2	2	1	1
1	–	–	–	–	–
3	3	3	2	3	–
2	1	3	2	1	–
3	1	2	1	1	1
3	3	1	–	–	–
3	–	–	2	1	–
3	1	3	2	2	1
1	–	3	–	–	–
3	–	–	3	2	–
–	1	3	–	–	–
2	1	2	–	–	–
2	–	2	2	1	1
1	–	–	2	1	1
1	2	1	2	2	1
1	3	1	1	1	–
1	3	–	1	1	2
1	1	1	1	1	–
1	3?	–	2	1	1
3	1	3	2	1	1
2	1	3	1	1	1
3	–	–	–	–	–
1	1	1	1	1	1
1	–	1	1	1	1
1	1	1	1	1	1
3	3	3	3	2	1
1	3	1	1	1	1
3	–	–	1	1	1
3	1	–	2	1	–
3	–	–	1	1	1

1 = satisfactory at 20°C (68°F); 2 = slight attack; 3 = attacked; – = no information.

Pyroceram[42] is a lithium aluminium silicate glass ceramic. This material is highly resistant to chemical attack, is non-porous and will withstand temperatures in excess of 500°C (932°F). It also has very high scratch and impact resistance.

Disposable coverings, such as polythene or polyvinylchloride sheet (sometimes bonded to an absorbent layer) and bitumen-interleaved sisal paper, are sometimes used as temporary protectives. They may be used to protect permanent surfaces from chemical attack or for convenience in confining a contaminant, but may present problems of hazard and of bulk when being removed for disposal.[12, 26]

3.4 Laboratory furniture and fittings

The laboratory furniture should be designed and constructed so that any contamination can be removed easily, and it must be so arranged that it does not impede egress in an emergency.

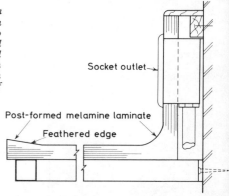

Figure 3.2 Section through a bench-top with melamine resin plastics laminate formed to provide an integral upstand 150 mm high at the wall and raised along the front edge of the bench to a height of 5 mm to form a dish to contain liquids.[13] *(Courtesy of Koch-Light Laboratories Ltd)*

Cracks and joints should be avoided in a working surface. It is an advantage if the bench-top can be formed into a shallow tray, say 5 mm deep, to contain spillages and also be provided with an integral or carefully attached upstand where it backs on to a wall. Several of the materials listed in Section 3.3 can be formed in this way, and the material chosen will be a compromise based on the various properties and, perhaps, the cost.

Basic Design Features

Figure 3.3 Section through a bench-top with a flat melamine resin plastics laminate and a separate iroko upstand. Note that the upstand is screwed and sealed to the bench-top and not secured to the wall, so that there is no gap between the upstand and the working surface.[13] (Courtesy of Koch-Light Laboratories Ltd)

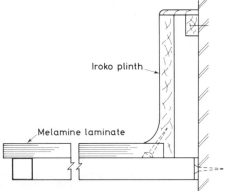

Several designs employing melamine resin plastics laminate[13, 29, 43] have been used successfully for bench-tops (*Figures 3.2–3.4*). One particularly convenient arrangement is shown in *Figures 3.5* and *3.6*. Here the bench-top for wet work is raised along all four edges to a height of, say, 5 mm with upstands attached along those edges which back on to walls. The working surface is then continued as a thicker flat bench-top where other runs of benching butt up to the first. This arrangement is advantageous in the corners of rooms because it avoids the difficulty of having to form on site an accurately mitred joint between two post-formed bench-tops of the style shown in *Figure 3.2*.

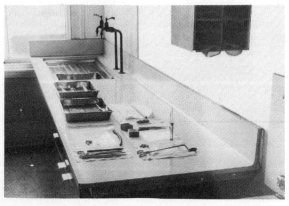

Figure 3.4 Bench-top made by post-forming a melamine resin plastics laminate (cf. Figure 3.2)

Figure 3.5 Section through a bench-top for wet work raised along all four edges to form a dish. Upstands are attached along those edges which back onto walls and the working surface continues as a thicker flat bench-top where other runs of benching butt up to it. (As to be used in the new Medical School Building, The University of Leeds)

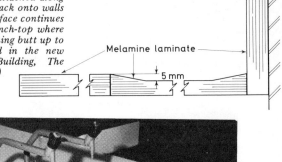

Figure 3.6 (top) Bench-top for wet work with all four edges raised to form a dish and fitted with a stainless steel sink bonded to the underside of the melamine laminate; (bottom) view of the underside of the bench showing the stainless steel sink bonded directly to the melamine laminate with a thermosetting or other type of resin. (Prototype of benching proposed for the new Medical School Building, the University of Leeds)

Basic Design Features

Similar designs can be effected using the plastics sheet materials or the stainless steel listed in Section 3.3. Here the material can be taken over the 5 mm lipping and finished off by turning under the bench-top.

The sinks in laboratories must be designed and installed so that they do not harbour contamination. It is generally preferable for the sink in which contaminated apparatus will be washed to have an integral draining board with a generous upstand of at least 25 mm all round. A stand-pipe overflow is less likely to retain contamination than a weir type, as it is more easily cleaned.

Selected stainless steel is an attractive material to use for integral sink and drainer units, but its susceptibility to attack by certain chemicals — in particular, hydrochloric acid — must be considered (*Table 3.3*). Alternative materials, e.g. polypropylene or earthenware, may have to be used for a proportion of the sinks in a laboratory. It is, however, very difficult to make a hygenic joint if an earthenware sink is to be set into a bench-top, and such a sink may be better installed as a free-standing item. If stainless steel is used, the sink and drainer unit can be sealed into a bench top in a number of ways, e.g. the stainless steel can be extended as a horizontal flange which is screwed to the underside of the bench top at 100 mm centres, compressing a sealant introduced between the flange and the bench-top (*Figure 3.4*), or, if the working surface of the bench is a melamine resin laminate, a very good seal can be formed by bonding the stainless steel to the underside of the melamine laminate *itself* with a thermosetting or other type of resin (*Figure 3.6*).

Sinks should be provided with elbow- or foot-operated taps to reduce the risk of transferring contamination from the hands (*Figure 3.4*). Taps can be provided which have screwed hose union outlets to enable water ejector pumps to be attached securely where the use of such pumps is permitted by the water supply authority. Where splash-backs are fitted, they should be of one-piece construction, e.g. of Vitrolite, and not tiled. Plastics or plastics-coated metal peg draining racks are preferable to absorbent wooden ones.

Basins specially for handwashing must be provided; if the laboratory suite has a changing room, these wash basins should be on the 'dirty' side of the change barrier with preferably an extra basin on the clean side to deal with any

Basic Design Features

contamination not removed by the first wash.[16] Disposable paper towels are preferable to laundered towels and should be treated as contaminated waste; they can be collected in a polythene bag mounted on a bag holder which incorporates a heating element to seal the bag when full (*Figure 3.7*). Plastics or rubber gloves can be kept in a divided storage unit, which forms 'pigeon-holes' each about 120 × 120 × 300 mm deep, made of wood finished with hard gloss or strippable paint (*Figure 3.4*).

The waste pipes carrying hazardous materials from sinks should be clearly labelled and sealed throughout their run to any treatment plant provided or to a relatively large drain where there is a sufficient quantity of uncontaminated water to provide a high dilution factor. The material chosen for the waste pipes is often high-density polythene or polypropylene

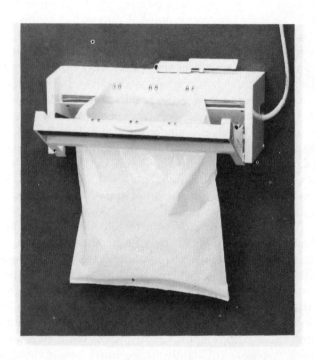

Figure 3.7 Waste disposal unit which incorporates a heating element to seal the polythene bag[44]

or glass; care must be taken to ensure that plastics pipes are adequately supported and that glass pipes are properly guarded against mechanical damage. Also, provision must be made for the thermal expansion of plastics pipework; at 20°C the coefficient of thermal expansion for polypropylene is 1.10×10^{-4} per deg C, which results in an increase in length of 7.2 mm in a 6 m long pipe per 10°C rise.[7] Small-volume running-traps are preferred to the larger-volume catch-pots unless there is a specific need for the latter; if catch-pots are provided, it may be preferable for them to be made of glass even in a mainly plastics system, because softening has been known to occur in plastics catch-pots at the level of the liquid surface.

The route taken for the drainage system through a building should be chosen with regard to the difficulties that would arise if a leak developed with consequent loss of containment of contaminated liquid. Waste pipes should not be installed close to water supply pipework, either buried or in ducts; where both are contained within a horizontal duct, the waste pipe should be below the water supply pipe.[7] The provision of adequately steep gradients is important to prevent blockages, especially in laboratories where ion-exchange resins are used or where macerators are installed. Also, the layout of the drainage system must allow for ease of rodding to free any blockages that do occur.

A possible hazard has been described involving the chemical sodium azide.[45] This chemical can react with metal drains to form deposits which are explosive. Heavy metal azides, particularly copper azide, are very shock-sensitive and deposits could be detonated by the mechanical action of clearing blockages by rodding. At particular risk are haematological and other pathological laboratories because of their use of sodium azide. The material is, however, easily neutralised, and this should be done by the user before discharge in order to avoid the hazard. Also, the use of a glass or plastics drainage system obviates the problem.

3.5 Services

As far as practicable, items associated with the main services which require frequent attention by the maintenance staffs should be located outside the main laboratory area. It is also an advantage to group services where appropriate so that they enter the main laboratory through a demountable wall or ceiling panel designed so that additional services can be introduced at a later date without holes having to be drilled through brickwork, etc. (see p. 12). Service pipe-lines should be colour-coded [46] and the hand-wheels of control valves identified by shape.[47]

High general standards of design, installation and maintenance are necessary.[214] Some indications of the extent of the services required are given in a number of general publications,[2-4, 6-9, 14] and the advantage of using service floors between laboratory floors have been discussed.[4] The generation of unacceptable noise is a problem which requires careful consideration in designing and installing mechanical services.[48-50] In addition, the following points should be noted.

3.5.1 Lighting

Fluorescent tubes are generally required for laboratory lighting; there is some advantage in using a simple enclosed fitting, such as the Crompton Clenelite (*Figure 3.8*), which can be washed, or a hospital-approved type with a sloping cover to minimise the accumulation of dust and contamination.[6] In certain locations flameproof fittings may be required.[52] Tungsten lighting has been recommended for radioactive counting rooms.[53] A system of emergency lighting should be provided and in some instances it may be an essential or a legal requirement.

3.5.2 Electric power

In addition to the good practices required by the wiring regulations,[54] a labelled mains switch should be provided near the exit of the laboratory so that the electrical power outlets

Basic Design Features

can be switched off, independently of the lighting, in an emergency. In some installations miniature circuit-breakers have been used both for circuit protection and as the means of isolating the socket outlets.

3.5.3 Manufactured and natural gas

Gas supplies for burners using compressed air or oxygen, such as those on a glass-blowing bench, must be fitted with non-return valves because of the possibility of explosion if compressed air or oxygen is forced back into the supply.[55-58] The use of a pressure relief valve down-stream of the non-return valve is also required where the working pressure could exceed 275 000 Pa (40 lb/in^2 gauge).

If natural gas is used extensively — for example, in glass-blowing — a high standard of ventilation is required to remove the combustion products.[59]

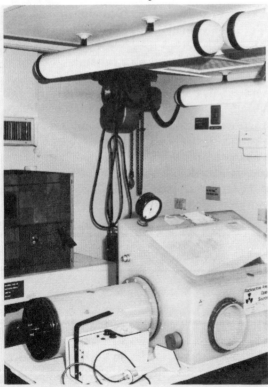

Figure 3.8 Washable enclosed fluorescent lighting fitting[51]

3.5.4 Laboratory gases

The decision to install a piped supply of any laboratory gas is usually taken on a balance of economic and safety factors. The economic factors are specific to the installation and are not discussed here. Several of the safety factors have been discussed elsewhere (Section 9.5).

The possible consequences of a faulty valve or of a leaking joint must be considered. In the case of a gas with a marked odour, such as ammonia, which can be detected by smell at a concentration just below the threshold limit value for 8 h exposure (Appendix C), a leak is rapidly detected and can be dealt with. A gas such as pure carbon monoxide or nitrogen or oxygen cannot be detected by smell, because it is odourless. If a leak of nitrogen, for example, continued for long enough, the oxygen concentration in the air in a confined space could be depressed to a point at which there would be a danger to life. Should an oxygen leak go undetected, then a very serious enhancement of the fire danger would result. Air normally contains 21% oxygen; if the oxygen concentration is raised to 24%, then fabrics such as wool, which only smoulder in air, inflame and burn rapidly when ignited.

When piped supplies of odourless gas are installed, the possibility of providing detectors and alarms should always be considered; unfortunately, suitable devices are not readily available for all gases in this category.

When piped supplies of fuel gases are installed, non-return valves and pressure relief valves should be fitted as for manufactured and natural gas (see Section 3.5.3) if compressed air or oxygen is also being used.

3.5.5 Steam supplies

The provision of steam supplies should normally be for fixed items of equipment. Pressure relief valves must vent in a safe manner, either to the open air or in an air extract duct. Appropriate thermal insulation of the pipes is essential.

Basic Design Features

3.5.6 Compressed air and vacuum lines

Compressed air lines create few difficulties. Good connections are essential and suitable patent couplings are advised (e.g. Schrader-type) in which a spring-loaded locking fitting is used.

Attention must be paid to the quality of the air supplied. The air intake for the compressor must be sited in a clean area, great care being taken to avoid places where fumes or gases are discharged or may percolate. If a compressed air supply for breathing purposes is required, it is recommended that a separate high-quality installation be provided; special provision will be required for removing oil droplets from the supply.

Extensive vacuum systems are useful only for comparatively soft (low) vacuums. Where a hard vacuum is required, individual pumps are generally found to be preferable. Vacuum can be lost in a common system by inconsiderate users or faulty valves introducing leaks into the system. It should be noted that vacuum systems are much more exposed to corrosion than are compressed-air systems. The exposure is to some extent controlled by the efficacy of the traps built into the system, but it is impossible to prevent the collection of a wide variety of chemical artefacts in the pump. In chemistry laboratories it is therefore usual to find that individual pumps are preferred. In laboratories using biological materials or radioactive materials the dangers of having a common vacuum system are obvious. The provision of a common system is therefore not to be recommended as a standard feature.

3.5.7 Lifts

In addition to the convential safety features of passenger and goods lifts, the following points should be considered.

1. Goods lifts in which dangerous materials are to be carried should be provided with an emergency telephone and alarm, so that, in the event of either a lift breakdown or an emergency in the lift-car, assistance can be obtained immediately and verbal communication established — presupposing the occupants of the lift are conscious.

Basic Design Features

2. The floor of the lift car should be able to retain spilled liquid and have reasonable chemical resistance. Rubber or PVC should be adequate for most purposes. The floor should be reasonably easy to swab down.
3. If a leak of dangerous gas, either heavier or lighter than air, takes place in the lift car, it should be possible to apply forced ventilation to the lift-shaft in order to carry out a rapid purge. It is important to be able to purge quickly the lighter gases which can collect at the top of the shaft and the heavier gases which can collect at the foot of the well or sump. As lift-shafts pass through several floors, any purge system must be by an extraction of air from the shaft, *not* by blowing air into it; otherwise the fumes will be spread into other parts of the building.

In multistorey buildings consideration must be given to the evacuation of casualties in an emergency. The height and shape of the building, the degree of compartmentation and the number of lifts in the various compartments will affect the final decision. At least one lift should be capable of taking a stretcher case and should be usable in an emergency situation. In very tall buildings at least two lifts which are approved by the local Fire Brigade for emergency use should be installed.

Lifts should be designed to enable disabled persons to use them easily. Manually operated doors are particularly irksome to persons in wheel-chairs.

When escalators or paternoster lifts are installed, conventional lifts are still necessary for moving goods and equipment and for use by people either temporarily or permanently disabled.

Four
Fire Precautions

Fire precautions in buildings constitute a specialised subject with a substantial literature, which is not easily summarised.[60-71, 212-214] Consequently, this chapter deals with its subject matter in a more generalised fashion than is the case in other parts of the book.

4.1 Consultation

The most important points to be observed are that statutory and local authority regulations must be complied with in the design of buildings and that the views of the insurance market should also be taken into consideration (see Appendix F). It is essential that at all stages of the project there be full consultation with the local Fire Brigade.* Consultation with the Fire Brigade cannot begin too early, e.g. advice obtained at the outset of the design stage may require the realigning or repositioning of a proposed building to provide adequate access for fire-fighting appliances. Furthermore, early and continuing close consultation ensures that the Fire Brigade officers are able to appreciate the

---

*When one negotiates with Fire Brigade officers, it may appear that over-severe standards are being set. This may well be so, and it is in order to question their reasoning. However, while doing so, they are entitled to forebearance. If there is a major fire, then the lives most likely to be at risk are those of their colleagues in the operations branch. Their knowledge of laboratory work is limited and is not reassuring to them. They are aware that corrosive or highly toxic vapours will be present and that gas cylinders and concentrations of highly flammable liquids constitute ever-present explosion hazards in a fire. The easier and safer it is for a fire-man to do his job, the less likely it is that a fire will get out of hand or that people will get hurt.

problems of design and operation of the building and are able to draw upon their fund of experience to advise on design details. The designer also learns to anticipate the probable reaction he may expect when considering alternative solutions to his problems.

The requirements of safety and security are often held to be incompatible. If the problems are faced at an early enough stage of the project, this is rarely, if ever, true. The basic requirements are simple and straightforward, but if they are ignored, needless difficulties for the users of the building can arise. It follows that the officers responsible for safety and security should be consulted by the designers before irrevocable decisions on the design of the building have been taken. These consultations should coincide with those involving the local Fire Brigade.

4.2 General principles

Fire precautions are aimed at protecting personnel and property against the effects of fire, and may be considered under two main headings: (1) passive measures, e.g. building design and layout, choice of materials of construction; and (2) active measures, e.g. choice of alarm systems, provision of fire-fighting equipment, administrative organisation.

Before we discuss the principal precautionary measures available, a brief note on the nature of fire is given.

Fire is a vapour-phase reaction for which three components must be present. These three components, often called the Fire Triangle (*Figure 4.1*), are: a supply of *fuel;* a source of *heat* sufficient to cause the fuel to give off a flammable vapour; and a support medium, usually the *oxygen* in the air, with which the fuel may react in the burning process. If any one of these three components is absent, a fire cannot start; if any one of them is removed from a fire, the fire will be extinguished. In the majority of instances a source of ignition is also necessary to initiate a fire when the three components are present but are below the auto-ignition temperature of the fuel. In a confined space, such as a laboratory, the ignition of a mixture of flammable vapour and air can take place explosively.

Fire Precautions

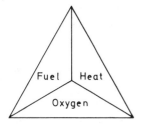

Figure 4.1 The Fire Triangle

The fire triangle is a useful concept to have in mind during the design of fire precautions. The role of fuel is obvious, and the potential fire hazard in a building is closely related to the fire loading, i.e. the amount of combustible material present in the building, and the nature of the combustible material. A piece of timber and an open container of petrol of equal calorific value do not present equal fire risks. The heat needed to raise a piece of timber to a temperature at which it will burn is substantial, although once alight with a good supply of air it will burn with vigour. On the other hand, organic liquids, similar to petroleum ether, which are often present in substantial amounts in laboratories, give off flammable vapours at well below room temperature, and are always high fire risks.

The support medium is usually the oxygen in the atmosphere, but it may also be present as cylinder oxygen or in certain oxygen-rich chemicals.

It must be remembered not only that oxygen is necessary to the fire but also that it is vital to people. In the presence of a fire, people are in competition with the fire for air to breath. That air must be cool and free from toxic or irritant gases and must contain sufficient oxygen not only to support life, but also to enable the person breathing it to act vigorously and to think clearly.

When a fire is burning in a confined space, there may be present gases too hot to breath safely, smoke and other irritant combustion products, and the highly toxic carbon monoxide resulting from incomplete combustion and an insufficiency of oxygen.

Fire precautions are therefore based on the need: (a) to prevent the conditions for a fire developing; (b) to limit the potential size of the fire, should one occur; (c) to prevent the fire spreading; and (d) to provide adequate safe means of egress for personnel. Fire fighting is based on: (a) cutting off the supply of fuel; (b) cutting off the supply of oxygen; or (c) cooling the fire to a temperature at which it can no longer sustain itself.

4.3 Building layout and emergency escape routes

Building layout must ensure that there are at least two distinct escape routes from the principal work areas, and from areas of high hazards, so that if one route is obstructed, e.g. by smoke, an alternative route is available.

In large buildings or where there are high fire risks, the buildings must be compartmented.[68, 223] The aim of compartmentation is to prevent the spread of fire, smoke and hot gases within a building. There should be provision for both vertical and horizontal containment of fire and combustion products. Heat and combustion products tend to travel upwards, but the water used in fire-fighting travels downwards and therefore the designer must aim to limit the damage which can follow as a consequence of both these facts.

The maximum size of the compartments will be specified in discussions with the local Fire Brigade. It should be borne in mind that most building regulations are written with places of public congregation, offices and similar places in mind. Laboratories contain hazards which are much more severe, so that prudence and common sense will suggest more rigorous standards of compartmentation than for ordinary buildings.

Escape routes must be usable by anyone who may have business in the building, at any time of the day or night. The term 'anyone' includes established members of staff, newcomers, visitors and strangers to the building, maintenance personnel and contractors' staff. The routes must be usable by disabled persons, whether they are temporarily disabled, e.g. a person with a leg or foot in plaster, or permanently disabled. It must not be assumed that a laboratory organisation will not include persons with one of the following disabilities: blindness, cerebral palsy (spasticity), muscular dystrophy, paralysis due to poliomyelitis, or other locomotor defects. Because of these considerations it is usual for Fire Brigades to insist that escape routes lead out of the

Fire Precautions

building via doors, and therefore windows are not accepted as recognised escape routes. Spiral staircases may also be frowned upon for the following reasons: first, a rapid descent can induce giddiness; second, only the outer part of the tread is wide enough for safe use; third, in the absence of landings there are no resting places or places where anyone falling may have their descent arrested; fourth, it is easier to move casualties down a zig-zag staircase with landings than down a spiral. Open-air fire escapes must be safe to use in frosty or snowy weather.

It is rare for large buildings to be completely deserted. Outside normal working hours, security patrols, maintenance staff and cleaners as well as the occasional member of staff dealing with an emergency or carrying out lengthy experiments over an extended time-scale may be present. Where there is animal accommodation, it follows that during holidays staff will be visiting the animals.

The ventilation of corridors, foyers and stairwells is important; the design should, first, prevent smoke spilling into these areas and, second, ensure that any smoke which does enter can readily escape to atmosphere. Doors opening into circulation areas should be fire resisting and smoke-stop doors should be provided where corridors join foyers.

It is important that goods entrances and stores or postal despatch and receipt offices should not be thoroughfares in the emergency escape system. It is very desirable that stores areas should be considered as separate entities in large buildings and be designed as fire compartments.

Human nature is such that if emergency exits are more conveniently sited than the normal exits, staff will find ways and means of overriding the locks and security may be lost. Consequently, the officers responsible for safety and security must agree and approve the emergency exit arrangements and the locking system to be used at an early stage in the design process.

Corridors, stairwells and foyers must be provided with sufficient emergency lighting to enable all persons present in the building to escape in the event of a failure of the main lighting. Exits must be clearly marked and, in deep-structure buildings particularly, exit routes must be signposted.

Fire Precautions
4.4 Fire alarms and fire detectors

The fire alarms and fire detectors may be only a small part of the emergency warning system provided in a large laboratory building (and indeed may not be the largest part of the system). The control system for the various services in the building may include a data print-out system for all the services, and it may therefore be expedient to link the fire detectors and alarms, as well as any security alarms, into this system.

There are four main functions of a detection and alarm system; (1) to detect malfunction of equipment, fire or other hazard; (2) to alert staff to the malfunction or hazard; (3) to control or suppress the hazard; and (4) to enable staff to move to a place of safety.

When the specification of a building is being drawn up, it may not be necessary for all these facilities to be included in the specification. However, for most buildings items (2) and (4) will be required to satisfy the local Fire Brigade. In large buildings some areas may require more elaborate precautions than others.

4.4.1 Preparation of the specification

A clear policy directive on fire and other emergencies must be given by the client. The designer of a detector and alarm system must have clear instructions on the following points.

Alarms
- (a) When is an alarm to operate?
- (b) Who is to be alerted by the alarm?
- (c) What is the alarm to mean, e.g. is the alarm to be an alert or a signal to evacuate the building? In the former case, is a second distinctive signal needed for evacuation of the building?
- (d) Is a manually operated or an automatic alarm required?
- (e) Is the alarm signal to be aural or visual or both?
- (f) What action should follow the alarm?
- (g) How many types of alarm are required?
- (h) Is an all-clear signal required?

Fire Precautions

If the signal is to be aural, there are three clearly distinguishable sounds to choose from: (1) bells, (2) sirens, (3) klaxons. These may sound intermittently or continuously.

If a visual alarm is specified, it may be a flashing light or a steady light. Because approximately one male in ten is wholly or partially red-green colour-blind, the use of colour alone is to be avoided; symbols, words or shapes should be introduced which readily convey meaning.

Although there is no officially recommended standard, the following is a very useful aural system: (a) bells, to indicate equipment failure; (b) sirens, to indicate fire; (c) klaxons, to indicate special hazards.

Bells seem to convey less urgency than either sirens or klaxons, and emit a sound markedly less penetrating than that from sirens. Sirens are more likely to be heard in the difficult nooks and crannies of the building.

Detectors

The specification must indicate: (a) whether automatic detectors are required, and, if so, what it is that must be detected; (b) whether the detectors should automatically operate the main alarm system or merely give a local warning; (c) whether a built-in fire-extinguisher system is required and if so whether it should be automatically operated by the detectors or be manually operated.

4.4.2 Detector systems

Detectors may be obtained for many things, e.g. fire, ionising radiations, burglars, toxic gases, services failure or equipment malfunction.

Fire detection. The problem of fire detection should be considered in two parts: first, for normal working hours, when the building is fully occupied; and, second, for the so-called 'silent hours', when the major part of the building is either unoccupied or sparsely populated.

During normal working hours, the staff themselves are the most sensitive part of the fire detection system. There is no detector as efficient as the eyes and nose of an alert human being.

Fire Precautions

The majority of fires start as small outbreaks and can be dealt with promptly, with portable extinguishers, before they have the chance to develop into a major incident.

Many serious fires start in areas which are unattended. This points to the advantages to be gained from the use of automatic detectors in selected areas, either coupled to an alarm system alone or, additionally, to an automatic extinguisher system.

4.5 Computers

Although large computers are usually housed in separate, specially designed buildings which are outside the scope of this book, there is increasing use of small or medium-sized computers in laboratories. Attention is drawn to a Department of Trade and Industry publication, *Computer Installations: Accommodation and Fire Precautions*.[70]

4.6 Problems associated with flammable and toxic solvents and with explosives

Organic solvents constitute one of the major sources of hazard in laboratories. They all, to a greater or lesser degree, give off toxic vapours, and the majority are highly volatile. Most flammable solvents may, as a rough rule of thumb, be regarded as being the equivalent of motor spirit. The non-flammable solvents may act as fire suppressants in some circumstances, but in a major conflagration the fire may be of sufficient intensity to cause them to decompose thermally, producing corrosive vapours without materially suppressing the fire. In any case the non-flammable solvents would produce large volumes of dangerous vapours.

It is therefore essential that adequate fire-resistant solvent storage be provided.

The storage of flammable solvents is subject to several statutory constraints: the Petroleum (Consolidation) Act 1928; the Petroleum, Inflammable Liquids and other Dangerous Substances Order 1947; the Petroleum Inflammable Liquids Order 1968; the Highly Flammable Liquids and Liquid Petroleum Gas Regulations 1972.[68]

Fire Precautions

Local Authority Fire Brigades are empowered to inspect premises and to issue licences for the storage of flammable solvents. Advice is given on the design of solvent storage units, and approval of designs should always be obtained before ordering materials (see Section 9.2).

The disposal of waste organic solvents poses many problems. The disposal method must comply with the requirements of the Disposal of Poisonous Wastes Act 1972* and with the Public Health Acts. If the solvents are incinerated, then the provisions of the Clean Air Acts must also be met.[68]

Quite apart from the legal constraints, it is highly dangerous to pour solvents into a drainage system. This practice can lead to explosions or fires in drains and/or the release of toxic vapours in parts of the building remote from the original disposal point.

There is a growing tendency for solvents to be supplied in large metal containers, e.g. of 5, 10 or 40 gallon capacity. This means that the main store must be associated with a solvent dispensary where the solvent can be safely dispensed into smaller containers for laboratory issue. This dispensary must be well ventilated and be free from sources of ignition. Flame-proof electrical fittings are strongly advised. Great care must be taken to ensure that spilled solvent cannot get into drains (see Chapter 9).

If explosives are to be used in experimental work, then the requirements of the Explosives Acts[68] must be met. Both the Police and the Fire Brigade must be satisfied that the arrangements for the storage of classified explosives is adequate before they will issue a licence for their being kept.

---

*Superseded by the Control of Pollution Act 1974.

Five
Means of Detecting and Extinguishing Fires

5.1 Automatic fire detectors

Automatic fire detectors are normally installed by specialist firms, and these notes are mainly intended as a short guide to the types of detector available and the general principles on which they operate. The following are the usual types of detector from which a choice will be made.

Combustion gas detectors (Figure 5.1). The detector head contains an ionisation chamber which responds to the gaseous products of combustion. The chamber is in balance with a similar but closed chamber. When the balance is upset, a cold cathode trigger tube is actuated and sets off the alarm.

Smoke detectors (Figure 5.2). Under normal conditions relatively clear air is present in the detection chamber. A beam of light shines across the chamber on to photocell A and no appreciable light reaches photocell B. When smoke particles enter the cell, light is scattered and some of the light falls on to photocell B, thus causing an alarm circuit to be triggered.

Heat detectors. There are several sorts of heat detector available:

1. Fusible solids are used in one of the simplest types of fixed-temperature heat detector, e.g. a fusible link damper in a ventilation duct (see Section 7.4).
2. Expanding liquid detectors can be used as either line detectors or point detectors. A line detector is designed as a long thin sensor, while a point detector is small and approximately circular in shape. This

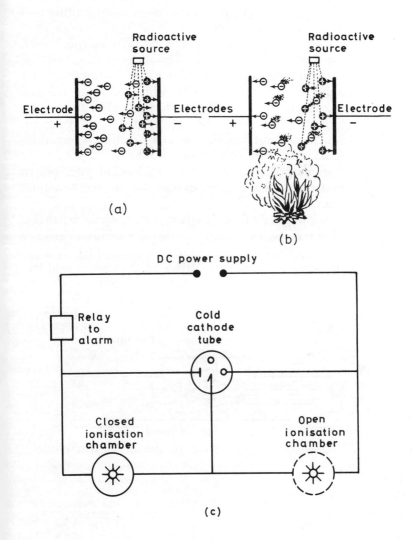

Figure 5.1 Combustion gas detector: (a) open chamber in the absence of combustion products; (b) open chamber with the ionisation current reduced by recombination of ions due to the presence of combustion products; (c) schematic circuit diagram. (Courtesy of the Fire Protection Association and Dunford Fire Protection Services Ltd)

type of detector can be arranged to react to a rapid rate of temperature increase or to slow heating to a high temperature. A thin metal tube containing a liquid is connected to two metal bellows which each support a contact arm. The flow to one of the bellows is restricted by a very small inlet, so that a rapid rise in temperature will cause an unequal movement of the contacts, thus actuating the circuit. Slow temperature changes result in an even expansion which does not actuate the circuit until a temperature substantially above normal is reached.

3. Pneumatic detectors work on a similar principle to that of the liquid type, except that the working fluid is air.
4. Expanding metal detectors use the thermal expansion of the metal to actuate alarms. They can be designed as fixed-temperature sensors or as rate of temperature rise detectors using the differential expansion of two dissimilar metals or a metal and non-metal combination.

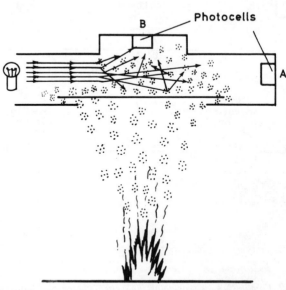

Figure 5.2 Smoke detector (optical). Light scattered owing to the presence of smoke particles in the detector reaches photocell B instead of photocell A and causes the circuit containing the photocells to become unbalanced, triggering the alarm

Means of Detecting and Extinguishing Fires

5. Thermocouple detectors can be used to detect either convective or radiant heat sources.
6. Detectors based on materials whose electrical properties change with temperature can be used as either line or point detectors.

Flame detectors. Flames can be detected by sensing the infra-red or ultra-violet radiation emitted. To avoid false alarms from radiant sources other than fires, the detector must be designed to respond only to the flicker frequencies emitted by flames, 5-30 Hz, or by selective sensing of chosen frequencies.*

Flammable vapour detectors. The sensitive element in the detector has catalytic properties which, in the presence of flammable vapours, cause a chemical reaction to take place. The heat of reaction produces a change of temperature which alters the electrical resistance of the element. The element is incorporated in a bridge circuit which becomes unbalanced when the resistance changes, causing the alarm to be actuated.

Laser-based detectors. A special form of optical detector which is under development.

Ultrasonic detectors. Ultrasonic detectors are based on the setting up of an ultrasonic wave pattern within a room. Convection currents from a fire cause variations in the air density, which in turn alter the velocity of sound in air, causing the standing wave pattern to break up and the alarm to be actuated. This type of detector is likely to be of very limited use in laboratories where convection currents set up by apparatus and experiments could cause false alarms.

5.2 Installation of detectors

A high standard of reliability is required of an alarm system.[71] Two sources of power are normally recommended, particularly for large buildings, e.g. the electrical mains with

---

*In an installation known to the authors several false alarms were experienced with a flame detector set in a large glass roof-dome which was surrounded by an area of flat roof. After rain, puddles of water collected on the roof and sunshine could be reflected from these puddles, through the glass windows in the dome, on to the flame detector. When there was a gust of wind, ripples on the puddles caused a flickering effect in the reflection which set off the alarms.

stand-by batteries for emergency use should the mains fail for any reason. When a detector system is chosen, it is important to ensure that adequate arrangements for inspection and maintenance are available.

In large buildings the system should be zoned and a central enunciator board provided to enable the source of the alarm to be located easily. The zoning should be by floors or parts of floors, unless there are very compelling reasons for vertical zones. The advice and agreement of the local Fire Brigade should be obtained on plans for the system. The enunciator board should bear an easily recognised relationship to the layout of the buildings.

The actual location of individual detectors should take into account the performance characteristics of the detector head (see Fire Data Sheet FS6005)[67] and the danger of physical damage by, for example, collisions with high loads on fork-lift trucks or trolleys. The equipment should be out of reach of casual tampering.

Alarm buttons should not be placed where tired members of the community can lean against them and set off the alarms, nor should they be where a clumsy cleaner can accidentally push a mop handle through a break-glass unit.

Before automatic drench systems linked to automatic detectors are proposed, considerable thought should be given to the consequences of false alarms actuating the drench system. The risk of a fire not being attacked quickly enough in an unlinked system must be balanced against the damage which might result from a false alarm setting off the automatic drench system automatically. Where an automatic fire extinguisher system is linked to the detector array, the risk of damage due to false alarms can be reduced by installing circuits which depend on more than one detector being activated. This system carries risks if the detectors are widely spaced.

5.3 Fire extinguishers*

The provision of an adequate supply of suitable fire extinguishers must be regarded as an essential part of the

---

*All fire extinguishers gain in effectiveness when used by trained personnel. The training of staff in the use of the equipment provided is therefore very important.

Means of Detecting and Extinguishing Fires

design and furnishing of a laboratory complex. Normally, the officer responsible for safety and fire precautions and the local Fire Brigade will indicate what is to be provided. The equipment falls into two classes — permanently installed items and portable items. It is essential that a schedule of this equipment be drawn up at an early stage in the design so that when detailed designing of laboratory layouts and furnishing commences, sufficient wall space is provided in suitable locations for the mounting of the fire-fighting equipment.

Generally, fire extinguishers in laboratories will be placed near doorways, so that, if they have to be used, the person concerned will move towards an exit when collecting the extinguisher, thereby putting himself between the fire and line of safe retreat.

Installed fire extinguishing equipment. Fixed installations include dry rising (water) mains; wet rising (water) mains; hose reels; sprinkler systems; mist sprays; foam inlets; carbon dioxide or vaporising liquid drench systems; and automatic powder extinguishers.

Portable fire fighting equipment. The following items are available and the choice will depend on local circumstances: woven glass fibre blankets; woven asbestos blankets; carbon dioxide extinguishers, ranging from 2½ lb (1 kg) capacity upwards (5 lb, 7 lb, 10 lb); carbon dioxide expelled water extinguishers (standard size, 2 gallons); pressurised water extinguishers (standard size, 2 gallons); foam extinguishers (standard size, 2 gallons); powder extinguishers (from 3 lb capacity upwards); and vaporising liquid extinguishers. Carbon dioxide extinguishers of 10 lb capacity are the largest wall-mounted ones that can be comfortably handled. They have a gross weight of about 40 lb and are rather heavy for slightly built ladies to use. Larger sizes are available mounted on trolleys.

A variety of foams is available and the choice will depend on the type of fire expected. Similarly, several types of powder are available, the choice depending again on the type of fire expected.

5.4 Classification of fires, methods of extinguishing and choice of extinguisher

5.4.1 Classification of fires

The classification of fires best known by long usage is:

Class A Ordinary combustibles, e.g. wood, paper, cloth, plastics.
Class B Flammable liquid fires, e.g. petrol, paraffin, oil, cooking fat.
Class C Electrical fires.
Class D Metal fires.

A new European standard has recently been issued — Classification of Fires, BS-EN2: 1972.[69] This classifies fires as follows:

Class A Fires involving solid materials, usually of an organic nature in which combustion normally takes place with the formation of glowing embers.
Class B Fires involving liquids or liquefiable solids.
Class C Fires involving gases.
Class D Fires involving metals.

It will be seen that Classes A, B and D are unchanged. However, Class C now has a totally new meaning. The elimination of a classification for electrical fires seems to be a retrograde step, as a wide range of portable extinguishers are dangerous in the presence of live electrical circuits.

So far as the authors are aware, the old system is still adhered to in American publications.

5.4.2 Extinguishing methods

Extinction by cooling. Extinguishers which operate by cooling use water as the cooling agent. Some other extinguishing agents have a cooling effect but this is secondary to their main effect. Water can be applied as a jet, as a spray or as a mist. In the third instance the water mist extinguishes by cutting off the supply of oxygen, with cooling as a secondary effect. Water used for fire-fighting has the serious drawback that it can cause damage by flooding and by entering electrical installations.

Extinction by restriction of fuel supply. In laboratories this type of extinction is achieved, first, by restriction of the quantity of highly flammable liquid present to a minimum by managerial sanctions, and, second, by ensuring that main stop valves on supplies of flammable gases are readily accessible to fire-fighters. Although in this context it is a source of heat and ignition rather than a fuel, the main electrical supply should also be readily controlled by fire-fighters.

Extinction by oxygen deprivation. There are several methods of depriving a fire of air: (1) by smothering the fire with a blanket of woven glass fibre or other non-flammable material; (2) by displacing the air with an inert gas or vapour, e.g. carbon dioxide or one of the halogenated hydrocarbons which are commercially available; (3) by covering the fire with a layer of suitable foam; (4) by covering the fire with a layer of suitable powder; (5) by blanketing the fire with a very fine water mist.

In laboratories the majority of fires start as very small outbreaks which, given the correct extinguishers, can be extinguished promptly. Consequently, as clean a method as possible is required as the first line defence. This is provided by a combination of the woven blanket and the carbon dioxide extinguisher.

Circumstances will dictate what other extinguishers are required, but in most laboratories they will be back-up equipment to deal with the more serious situations.

Foam extinguishers are normally portable. Although installed foam flooding inlets are available as standard items of equipment, they would be provided only in special circumstances, e.g. where large quantities of flammable solvent are being used in the experiments. There are several types of foam available and designers are advised to seek expert advice before making a choice. The same remarks apply to powder extinguishers as to foam extinguishers. It is an unfortunate fact that foam and powder extinguishing agents make a lot of mess. This is a strong point in favour of these extinguishers being manually operated: the cleaning-up operation following the activation of an automatic system using foam or powder by a false alarm could be extremely expensive.

The inert gases and vapours most commonly used are carbon dioxide and halogenated hydrocarbons such as BCF (bromochlorodifluoromethane). These agents can be used either in fixed systems or as portable extinguishers. When an inert gas flooding system is designed, adequate safety precautions are essential. When the concentration of oxygen in the atmosphere is depressed below 16%, then persons present begin to be at hazard, this being the concentration at which the effects of oxygen starvation begin to appear. In a total flooding system, death due to oxygen deprivation can be very rapid. There MUST be a time delay between the detection and alarm and the subsequent operation of the flooding system to enable the occupants of the area to escape. After the flooding system has operated there MUST be a clear warning that the atmosphere is unsafe to breathe.

The use of halogenated hydrocarbons on very hot fires can result in their decomposing to form corrosive vapours.

5.4.3 Reliability of automatic systems

The reliability of automatic systems, although good, is significantly less than 100%.[71] Therefore the specification and design must allow for the occasional false alarm. These may occur several times during the commissioning stage of the project.

5.4.4 The choice of extinguisher

Class A fires (both old and new classification)	Water is the best extinguishing agent but any of the others may be used.
Class B fires (both old and new classification)	These may be extinguished by a blanket, an inert gas extinguisher, a foam extinguisher or a powder extinguisher. A water spray or jet MUST NOT be used on this type of fire. The burning liquid floats on the water, spreads and flares up as

Means of Detecting and Extinguishing Fires

	steam entrains the flammable vapour. Water-mist extinguishers, however, are available for fish fryers and similar installations.
Class C (electrical, old classification)	An inert gas may be used to extinguish burning material after the power has been switched off. Powder may be used as a last resort but it is difficult to remove afterwards. Water and foam must not be used from portable equipment, as they are electrically conducting.
Class C (gases, new classification)	The first action is to turn off the supply. Subsequently, secondary fires should be prevented by use of the appropriate appliance. (Beware of explosive atmospheres building up.)
Class D (both old and new classification)	Special powders are recommended. Generally, all other agents react violently with hot metals.

Six
Laboratory Ventilation

6.1 Methods of providing ventilation

Ventilation is one of the most important and expensive provisions in a laboratory. The method of ventilation chosen depends on the toxicity of the substances to be used and on the nature of the experimental work to be undertaken. Five methods of extracting contaminants are considered in this section; in addition, laminar air-flow techniques for clean rooms and work stations are dealt with in Chapter 8.

The five methods are:

(1) general dilution ventilation, (2) local exhaust or spot ventilation, (3) partial enclosures, i.e. fume-cupboards, (4) special enclosures and (5) total enclosures, i.e. glove-boxes and safety cabinets.

6.2 General dilution ventilation

General dilution ventilation is designed to introduce uncontaminated air continuously into the laboratory and to mix it thoroughly with the contaminant being released. It is not suitable for use as the sole method of controlling highly toxic contaminants and is often an adjunct to the provision of fume-cupboards. The air extracted through fume-cupboards should be replaced by fresh air introduced in such a manner that the whole volume of the laboratory is purged; the number, the design and the siting of the air inlet grilles are consequently very important. Care should be taken to ensure that they are at locations remote from the extract points and that the air is introduced at an acceptable velocity

Laboratory Ventilation

with the worker between the inlet and the source of the contamination (see Section 6.7).

If a large fume-cupboard is placed in a small laboratory, the rate of general ventilation expressed as the number of room air-changes per hour may appear to be unacceptably high for the comfort of the occupants. Discomfort may not, in fact, occur if attention is given to the air distribution within the laboratory and especially to the velocity at which the fresh air is introduced into the laboratory. The highest air velocity which need be in the laboratory in the vicinity of a person is the 0.5 m/s (100 ft/min) (see Section 6.4) for the air as it enters the fume-cupboard.

The specification of the dilution ventilation in terms of the number of room air-changes produced per hour is imprecise. However, alternative methods of specifying the required ventilation rates do not necessarily result in more precise control of the situation. An attempt could be made, for example, to calculate the required extraction rate if the rate of evaporation of a liquid producing the contamination were known:[72,73]

$$\text{extraction rate} = \frac{2.4 \times 10^7 \times (\text{evaporation rate in kg/s})\, H}{(\text{molecular weight})(\text{threshold limit value in ppm})}\, m^3/s$$

where H (*Table 6.1*) is a safety factor introduced to allow for non-uniform distribution of the contaminant throughout the fresh air brought into the room, which in turn depends on the relative positions of the extract points, the source of contaminant, the worker and the inlet points (*Figure 6.1*).

If this formula is applied to benzene (high toxicity), for example, assuming good distribution (i.e. $H = 7$) in a laboratory of volume 50 m³, an evaporation rate of 0.1 ml/min would require 10 room air-changes/h in order to limit the concentration to the maximum allowable of 25 ppm (see Appendix C).

In the same laboratory with the number of room air-changes/h kept at 10 and $H = 7$, hydrogen sulphide could leak into the atmosphere at a rate of 20 mg/min (10 ml/min at NTP) before the threshold limit value of 10 ppm would be reached. The initial limit of detection by smell (that is, the limit before olfactory fatigue occurs) is 0.13 ppm, and this would be reached with a leak rate of 0.2 mg/min (0.1 ml/min at NTP).

Laboratory Ventilation

Table 6.1 VALUES FOR H, SAFETY FACTOR FOR DIFFERENT DISTRIBUTION CONDITIONS[72, 73]

Toxicity*	Poor distribution	Moderate distribution	Good distribution	Excellent distribution
Slight	7	4	3	2
Moderate	8	5	4	3
High	11	8	7	6

* For chemical toxins a semi-quantitative classification based on threshold limit values (see Appendix C) is: slight, threshold limit value >500 ppm by volume in air; moderate, threshold limit value 101–500 ppm; high, threshold limit value <101 ppm.

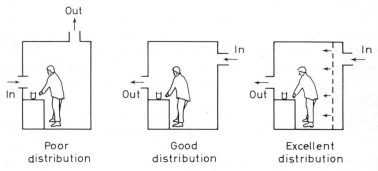

Figure 6.1 Poor, good and excellent distributions of the fresh air brought into the laboratory to protect the worker from the contaminant. Note the relative positions of the extract, the source of the contaminant, the worker and the air inlet. [72] *(Courtesy of* Chemistry in Britain*)*

Recommendations have appeared in the literature for the number of room air-changes per hour desirable in laboratories used for various types of work. Munce[1] gives rates of ventilation 'proved by experience' for chemical laboratories of 4-8 air-changes/h and, for rooms where obnoxious gases are produced, of 30. The Nuffield Foundation: Division for Architectural Studies[2] considers that, when most of the objectionable operations are carried out in a fume-cupboard, 8-10 air-changes/h 'may still be required in the laboratory itself'. For work with radioactive substances (see Appendix A) the International Atomic Energy Agency[53] recommends a minimum of 12. It does appear, therefore, that, in general, a laboratory should have about 12 air-changes/h or more of

Laboratory Ventilation

general ventilation. In some cases the fume-cupboards present may provide the required rate of general ventilation; if not, other extract points would be necessary and these would have to be integrated with the fume-cupboard extract system to preclude the possibility of reverse air-flow if a fume-cupboard were switched off.

6.3 Local exhaust or spot ventilation

To capture contaminants near their point of origin before they disperse through the laboratory is very attractive. Local exhaust systems can be applied for this purpose to equipment which cannot be readily enclosed, or because of the need for access, or because of brevity of use.

Although useful, this method of ventilation is not an alternative to the provision of a fume-cupboard if a relatively high degree of protection is needed. The drawing of air into a pipe is not an efficient way of capturing fumes; at a distance from the pipe equal to one pipe diameter, the velocity of the air drops to less than 10% of the velocity at the opening[74] and the scavenging effect is consequently not great.

Local exhaust terminations can be divided into at least three categories: (1) round openings or rectangular openings with width-to-length ratios greater than about 0.3, (2) slot openings with width-to-length ratios less than about 0.3 and (3) canopy hoods.

Approximate formulae exist for calculating the (volume) rates at which air must be extracted through a termination in order to remove contaminants released at different rates at various distances from the termination.[72,73,221] The formulae give more accurate answers if the velocity needed to capture the contaminant at source is known; if it is unknown, a table of recommended capture velocities can be used (*Table 6.2*). Note that many vapours are denser than air and consequently tend to collect at low level if there are no thermal rises or other disturbances present to counteract the tendency; this should be borne in mind when positioning a termination.

Round openings or rectangular openings with width-to-length ratios greater than about 0.3. These can be used where the contaminant arises from a physically small source (*Figure 6.2*).

Laboratory Ventilation

Table 6.2 RECOMMENDED CAPTURE VELOCITIES (m/s) FOR DIFFERENT RATES OF RELEASE OF CONTAMINANTS [72, 73]

Toxicity*	No upward velocity	Moderate upward velocity	Active generation	Violent† generation
Slight	0.25	0.4	0.6	2.5
Moderate	0.5	0.75	1.3	5.0
High	0.75	1.1	2.0	7.5

*See footnote to *Table 6.1*.
† Local exhaust ventilation is not recommended if the contaminant is violently generated.

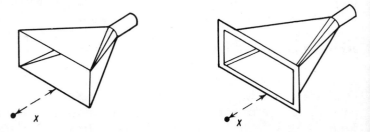

Figure 6.2 Rectangular openings without and with a flange. [72] *(Courtesy of Chemistry in Britain)*

The termination can be at a fixed position or fitted with a flexible extract duct to allow movement. The extraction rate in cubic metres per second required to produce a capture velocity V m/s at a distance X metres from a termination of cross-sectional area A m² is given approximately by:

$$\text{extraction rate} = V(10X^2 + A)$$

For example, to extract a substance of slight toxicity which is being generated with moderate upward velocity by a small source at 0.5 m distance from a termination of cross-section 0.5 m × 0.5 m, the extraction rate required is:

$$\text{extraction rate} = 0.4(10 \times 0.25 + 0.25)$$
$$= 1.1 \text{ m}^3/\text{s or } 2300 \text{ ft}^3/\text{min}$$

If it is possible to fit flanges around the opening, the capture efficiency increases and the extraction rate may be reduced by 25%. Gross-draughts greatly impair capture efficiency and, if any are present, side screens should be fitted.

Slot openings with width-to-length ratios less than about 0.3. These can be used where the contaminant arises from an extended source for which a long narrow exhaust termination, of length L m, is appropriate. The extraction rate required is given approximately by:

$$\text{extraction rate} = 3.7\ VXL\ \text{m}^3/\text{s}$$

Canopy hoods. Canopy hoods should be placed as near to the source of contaminant as possible; they are not recommended where there are strong cross-draughts unless side-screens are fitted. The extraction rate required is given approximately by:

$$\text{extraction rate} = 1.4\ PDU\ \text{m}^3/\text{s}$$

where P m is the perimeter of the base of the canopy, D m is the height of the base of the canopy above the top of the bench or tank and U m/s is the required velocity through the area between the edge of the canopy and the edge of the bench or tank.

Table 6.3 SUGGESTED VALUES FOR VELOCITY U (m/s) FOR CANOPIES [72,73]

Toxicity*	No cross-draughts	Slight cross-draughts (0.15–0.35 m/s)	Moderate cross-draughts (0.35–0.75 m/s)
Slight	0.4	0.6	0.9
Moderate	0.6	0.9	1.1
High	0.9	1.1	1.4

*See footnote to *Table 6.1.*

Suggested values for velocity U are given in *Table 6.3*. These may be reduced by 25% if screens are fitted on two opposite sides and by 50% if they are fitted on two adjacent sides, provided that U is not reduced below 0.25 m/s. Where there is a cross-draught, any screens fitted should be on the *leeward* side of the canopy.

6.4 Partial enclosures (fume-cupboards)[75]

The fume-cupboard is a useful form of partial enclosure provided that it is properly designed and installed (*Figure 6.3*). Here the term 'fume-cupboard' is used to mean a partially enclosed chamber with or without a sash fitted to the opening and provided with a mechanical extract system;

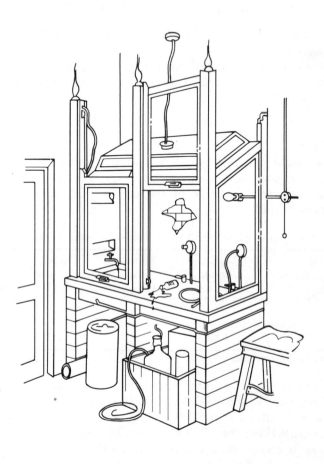

Figure 6.3 The fume-cupboard (circa 1928 — later in some institutions). Note the following design faults: (a) sited near door (cross-draughts at working face); (b) no aerodynamically shaped fascia (turbulent air-flow into cupboard); (c) no back baffle (poor scavenging of heavy vapours); (d) side extract (non-uniform air-flow across aperture); (e) no by-pass (excessive velocities as sash is lowered); (f) controls inside cupboard (operating hazard); (g) poor internal finishes (contamination risk); and (h) flat worktop (no retention of spilt liquids). (Drawing by courtesy of Mr Denys Horner, Senior Assistant Bursar (Planning), the University of Leeds)

Laboratory Ventilation

this is in contrast to the term 'fume-hood', which is reserved here to mean a canopy, perhaps with one or more sides, as described above (Section 6.3).

An official description of a fume-cupboard is given in the Ionising Radiations (Unsealed Radioactive Substances) Regulations 1968:[76]

' "a fume-cupboard" means a partial enclosure —
(a) having mechanical means of producing at any opening between it and the workplace a flow of air into it which has a velocity (being in any event a velocity not less than 50 cm per second) and is otherwise such as to prevent the spread of radioactive substances from the enclosure to the workplace;
and
(b) where the said flow of air is not kept in constant operation, provided with a shutter which is kept in its closed position when the flow of air is not in operation and which is such that when the shutter is in its closed position is a total enclosure.'

Although this definition occurs in UK Regulations relating to the use of radioactive substances in factories, none of the conditions in the definition arises specifically because the hazardous material is radioactive; the definition may therefore be taken as a guide to what should be provided in a fume-cupboard intended for use with other hazardous materials. The definition also requires that the fume-cupboard be capable of being converted into a *total enclosure* where the extract system is not operated continuously. Many commercially available fume-cupboards — in particular, those which are sashless and those which employ a simple by-pass arrangement to effect velocity compensation (see below), do not meet this second criterion.

The value specified for the minimum acceptable velocity of the air entering through *any* size to which the working aperture can be opened is also discussed below.

The actual design and the siting of a fume-cupboard must be such that the fumes generated within it are contained efficiently. This efficiency depends primarily on the velocity of the air moving into it through the working aperture and on the absence of eddy-formation in this airflow. The efficiency is affected by cross-currents in the air outside the fume-cupboard, the shape of the front of the fume-cupboard, the design of the extract slots, and obstructions, convection currents and mechanical action within the fume-cupboard.

The design and installation of a fume-cupboard system can be considered under three main headings: (1) the design of the fume-cupboard itself; (2) the siting of the fume-cupboard within the laboratory; and (3) the design of the extract ductwork, the fan and the fume dispersal system (see also Chapter 7).

6.4.1 The design of the fume-cupboard

The main requirement is an adequate velocity over the entire working aperture for the air flowing into the fume-cupboard. A velocity as low as 0.3 m/s (60 ft/min) will retain fumes in an aerodynamically styled (see below) fume-cupboard under *ideal* conditions. However, under *working* conditions a velocity of at least 0.5 m/s (100 ft/min) is necessary to ensure the capture and retention of the fumes against the effects of turbulence — in particular, turbulence due to movement of persons working at or walking past the fume-cupboard. This figure is in agreement with the minimum velocity of 0.5 m/s specified by the UK Department of Employment in the Ionising Radiations (Unsealed Radioactive Substances) Regulations 1968[76] quoted above (p. 57) and by the Department of Health and Social Security for fume-cupboards in pathology laboratories.[7] This figure is further substantiated in the report of the working party on fume-cupboard design of the British Occupational Hygiene Society,[77] which also recommends a test procedure for measuring air-flow.[78,79]

If an adequate air velocity is provided over the maximum working aperture of a fume-cupboard, excessive air velocities will result if the aperture is reduced by lowering a sash, unless some form of compensation is provided. There are several ways of limiting the air velocity; one of the easiest is to incorporate a simple by-pass which is uncovered as the sash is lowered (*Figure 6.4*). This method has the advantage that the rate of general ventilation of the laboratory is not appreciably altered when the sash is lowered, because the air entering through the by-pass is also taken from the laboratory. The method, however, does not meet the

Laboratory Ventilation

requirement of paragraph (b) of the extract from the Factory Regulations[76] quoted above (p. 57) that, if the air-flow is not kept in constant operation, the fume-cupboard must be converted into a total enclosure when it is closed.

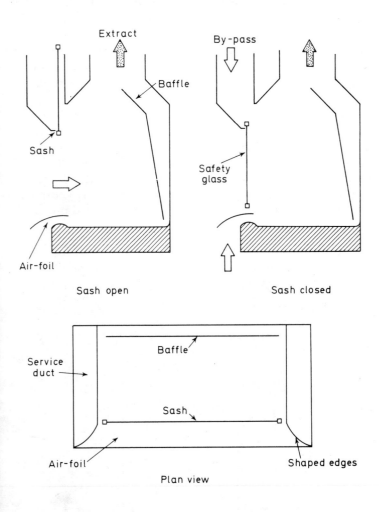

Figure 6.4 *An aerodynamic fume-cupboard with back baffle, air-foil, shaped edges and simple by-pass which prevents excessive air velocities when the sash is lowered*

Laboratory Ventilation

An alternative method of velocity compensation is illustrated in *Figure 6.5*.[80] Here the sash is coupled to a damper in the extract system so that the extract rate through the fume-cupboard is reduced as the sash is lowered. This arrangement can convert the fume-cupboard into a total enclosure when the sash is lowered and it maintains the level of general ventilation in the laboratory.

In order to avoid the formation of eddies, the front face of the fume-cupboard should be of an aerodynamic shape to produce a streamline air-flow. The placing of an air-foil along the front edge of the base to create a gap of about 25 mm between it and the base reduces eddy formation, improves containment in cross-draughts such as those produced by the operator[78] and ensures that heavy vapours are efficiently scavenged. A safety hump, rounded to match the underside of the air-foil, can be formed along the front edge of the base in order to contain spilt liquid; in some designs this also helps to ensure that apparatus is not placed too near the front of the fume-cupboard.

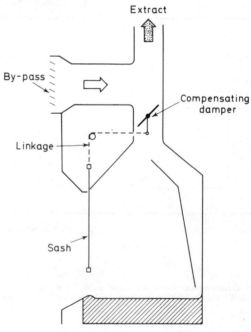

Figure 6.5 A fume-cupboard in which the sash is linked to a damper in the extract system in order to prevent excessive air velocities when the sash is lowered

Laboratory Ventilation

Turbulence within the fume-cupboard is reduced and the extraction efficiency, especially for heavy vapours, is increased if a back baffle is fitted to form a rear plenum with a low-level as well as a high-level extract slot (*Figure 6.4*). The baffle should be demountable for cleaning purposes; this is easier to achieve if the baffle can be divided into sections for removal. For fume-cupboards up to 1.2 m (4 ft) in internal width, one connection to the extract ductwork is usually sufficient; for wider cupboards, two connections are advisable in order to ensure reasonably uniform air-flow over the whole width.

In order to reduce the large volume of air required by a wide fume-cupboard, the effective width has been reduced in some designs by inserting one or more glazed screens to run in a *single* horizontal track across the front of the fume-cupboard.[95] Such screens may produce turbulence and should be investigated before use. Other designs, intended to reduce the amount of (conditioned) air taken from the laboratory, have auxiliary air supplies with fans which deliver untreated air directly into the fume-cupboard. These may constitute a hazard[77, 81, 82] unless the air-flows themselves are interlocked using air-flow sensors (see Section 7.3); if the extract system fails or becomes gradually unbalanced with respect to the system supplying the auxiliary air, the auxiliary air may blow fumes from the fume-cupboard into the laboratory. Also, in many cases completely untreated air is too cold and dirty to be introduced directly into the fume-cupboard.[7, 12, 77, 81, 82]

The fume-cupboards should be glazed in toughened glass or in other appropriate safety material and never in ordinary window glass.[77] Toughened glass is usually the material of choice; in most applications any superficial etching of the glass which might occur is unlikely to be deep enough to affect its strengh even over a long period of time, but in a large explosion the blunt grains of glass formed when toughened glass is shattered would be ejected. Laminated glass is designed to remain in position when it fractures, and consequently may be safer than toughened glass in a *moderate* explosion, but sharp-edged blisters may be projected from the outside surface; it has a poorer thermal shock-resistance than toughened glass and the plastics interlayer may weaken through chemical attack if organic vapours are present. A laminated, toughened, anti-explosion (and burglar-proof) glass is available but is expensive.

Wired glass provides protection against fire but is not suitable for fume-cupboard glazing, because splinters of the outer layer of glass may be projected in an explosion. In a fire the glass will crack, but it is held in position by the incorporated wire mesh. Chemical atmospheres may corrode the wire at the edges in time, causing splitting of the glass.

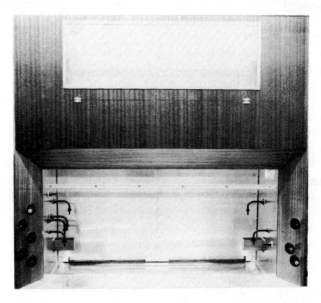

Figure 6.6 An aerodynamically styled fume-cupboard with air by-pass and with dished working surface and side-walls in FMB grade stainless steel[92]

Some transparent plastics materials are available with a higher impact resistance than that of glass. They are particularly useful when hydrogen fluoride, which etches glass, is present, but their temperature resistance and surface hardness are inferior to those of glass and some may shatter dangerously in an explosion.

Water and gas outlets should be sited towards the front of the fume-cupboard rather than at the back, so that they can be reached without it being necessary to put the head and shoulders inside the fume-cupboard and so that they are out of the main stream of any corrosive fumes present; their controls should be mounted outside the cupboard and are usually placed on the side facias, where they are clearly visible (*Figure 6.6*). Electrical outlets should be mounted low down on the side fascias so that there is less likelihood of cords to the appliances being pulled when the sash is lowered. Panels giving access to the plumbing and to the electrical wiring should not be fitted in the *interior* surfaces of the fume-cupboard for reasons of cleanliness; if the location of the fume-cupboard precludes the fitting of access panels in the exterior surfaces of the side walls, the provision of a demountable fascia should be considered.

The materials of construction for the base and the walls of a fume-cupboard are usually chosen from the lists in Section 3.3. Plastics materials are attractive because of their good chemical resistance, but chemical attack[83] can occur when acids are used (*Table 3.3*). Also, in the event of a fire, plastics decompose with the production of large volumes of smoke and, in some cases, with the production of very corrosive decomposition products, e.g. polyvinylchloride produces hydrochloric acid.[84-89,215] Stainless steel is sometimes chosen for both the working surface and the walls (*Figure 6.6*); when carefully welded, it gives a good finish and has good decontamination properties (*Table 3.2*), but it is relatively expensive and is susceptible to attack by some chemicals (*Table 3.3*). Aluminium treated with a suitable surface coating (see *Table 7.1* and Section 3.1) is often used for the walls; compressed asbestos and tiles are occasionally used for the working surface, but they are not generally suitable for work where cleanliness is important, because of the absorbency of the asbestos and the difficulty of cleaning between the tiles.

In radionuclide laboratories the fume-cupboard may have to be capable of supporting the weight of a lead shield erected on the working surface, and concrete shielding may be required as an integral part of the working surface (see Appendix A).

The provision of shelves in a fume-cupboard for the storage of hazardous chemicals should be discouraged. In addition to the possible adverse effects on the aerodynamics of the cupboard, a fume-cupboard is an enclosure in which a fire or a violent reaction may occur, and the presence of stocks of hazardous chemicals compounds the danger. Ventilated storage space can often be provided conveniently beneath a fume-cupboard. It is important, however, to ensure that fumes from such a storage cupboard do not come into contact with the service controls and pipework fitted to the fume-cupboard. The fumes should not be extracted directly into the fume-cupboard but should be introduced into the ductwork a little way along the extract system from the fume-cupboard for reasons of safety.

Special considerations apply if the fume-cupboard is to be used for work with hydrofluoric acid, which attacks glass, or with perchloric acid.[90,91] The vapours produced during perchloric acid digestions render materials such as wood highly flammable; also, many metals are attacked by the vapours to form crystals of metal perchlorates which are extremely shock-sensitive and violently explosive. Many serious fires, explosions and fatal accidents are recorded in the literature.[90,91]

Care must be taken with the design of a fume-cupboard for work with perchloric acid to ensure that no wood or unprotected metal surfaces are exposed to the vapour. It is also necessary to take similar precautions with the extract system, and no jointing compounds containing organic material may be used in case of violent oxidation reactions.

One design of fume-cupboard for work with perchloric acid[92] consists of a working chamber constructed of brickwork finished with either smooth concrete or tiles and with a rear baffle of plate-glass supported by plastics-coated lugs. The aerodynamic fascia is out of reach of the fumes and so is constructed from conventional materials. The extract system, including the fan, is made out of PVC.[93]

A fume-scrubber can be used to advantage to entrain the perchloric acid vapour and spray,[94] before it can enter the extract system.

6.4.2 The siting of the fume-cupboard

The fume-cupboard is one of the most exacting of laboratory fittings to site and it should be the first, rather than the last, item to be considered when the layout of a laboratory is planned (*Figure 2.2*). It must be recognised that the fume-cupboard will be the location of a large proportion of the hazardous experimental work carried out in the

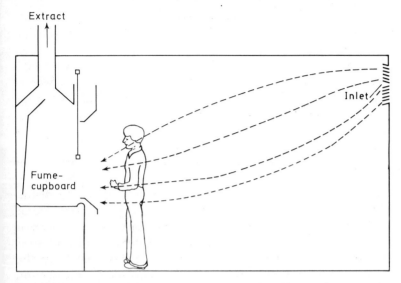

Figure 6.7 A room air inlet positioned to ensure that the air velocity near the operator's head is not excessive and that most of the room is ventilated

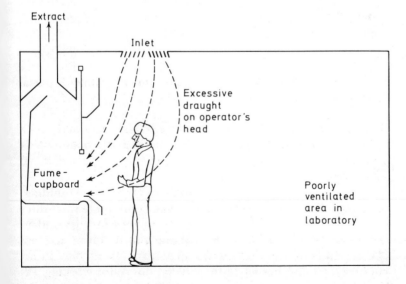

Figure 6.8 A badly placed air inlet causing excessive down-draughts and providing little general ventilation in the room

laboratory. The provision of large, efficient fume-cupboards is part of a trend towards the performing of the actual experiment in the fume-cupboard and the use of the open benches as places for assembling equipment and placing reagent bottles, spare glassware, notebooks, etc. There is therefore a strong argument for keeping the fume-cupboard as far from the entrance and the emergency exit as possible, so that in an accident the line of retreat is away from the hazard area.

As explained in Section 6.4 above on the design of fume-cupboards, the movement of staff working at or walking past the fume-cupboard sets up strong air disturbances which affect the performance of the cupboard. Major air disturbances are also caused by the opening and closing of doors. It is therefore prudent to reduce these air disturbances by placing the fume-cupboard as far as possible from the doorways and other busy areas in the laboratory. In addition, the air supplied to the laboratory should be via inlets (see Section 6.7) at the opposite end (or side) of the laboratory to the fume-cupboard; this ensures that (a) the air velocity near the operator's head is not excessive, (b) the whole of the room is ventilated and (c) the fume-cupboard can draw its air in a smooth stream perpendicular to the fume-cupboard face rather than with flow lines parallel to the face *(Figures 6.7, 6.8)*.

6.4.3 The extract ductwork, the fan and the fume-dispersal system

Contaminants having been captured and retained in a fume-cupboard, they must be conveyed in a safe manner to a point from which they can be effectively dispersed in an acceptably low concentration.

When there is more than one fume-cupboard in a building, the possibility of linking them into one common extract duct has to be considered because of the savings in cost and in duct-space that can be made. Among the disadvantages are (1) the possibility that fumes may drift from one fume-cupboard to another when the extract fan is not working, (2) the loss of personal control by the worker over the fume-cupboard that he is using, with the possibility that the

Laboratory Ventilation

extract may be turned off by others without prior warning and (3) the imbalance possible among the extraction rates from the individual fume-cupboards due to inaccurate commissioning or subsequent unauthorised adjustment of dampers. One important advantage of individual ducts is that a hazardous material released in one fume-cupboard is confined to one duct and not distributed more generally. For example, ducts containing radioactive substances are usually kept separate from others and a permit-to-work system operated for maintenance personnel.

If a linked system is employed, it is advisable to provide the operator of a fume-cupboard with a visual signal to indicate when the extract system is functioning correctly; such a signal should be actuated by an air-flow sensor (see Section 7.3) and not just be *electrically* interlocked with the fan motor.

The design of extract systems for fume-cupboards is discussed in further detail in Chapter 7.

6.5 Special enclosures

In some cases special enclosures[92] can be constructed around apparatus to prevent the dispersal of aerosols,[17] the accumulation of explosive mixtures of flammable vapours, etc. For example, they can contain the aerosols formed around a blender (*Figure 6.9*) or when tissues are aerated in radioactive solutions. Automatic fraction collectors can be placed in ventilated enclosures to ensure that no explosive mixture builds up which could be ignited by the electric motor which drives the collector,[95] and in some cases it may be necessary to ventilate animal cages individually.[6]

The effectiveness of special enclosures in preventing the dispersal of contaminants is governed by the same principles as those applying to fume-cupboards (see Section 6.4) or to other open-ended ventilation systems: (a) the velocity of the air entering through any opening must be sufficient to prevent reverse air currents from being set up in the opening which would allow of the escape of the contaminant and (b) the velocity of the air drawn into the extract ductwork must be high enough to entrain the contaminant if it is intended to convey it away from the enclosure. If the degree of containment required of a given special enclosure is critical, the design must be tested. Generally speaking, an air velocity of 0.5 m/s (100 ft/min) through any opening ought to be sufficient but, as the opening may not be designed to give streamlined air-flow, eddy-current formation may dictate a higher velocity.[7,96-98]

When ventilation systems are planned, it may be appropriate to provide in some laboratories spigots near bench level to facilitate and, hence, to encourage the connecting of special enclosures to the extract ductwork; such an arrangement might well be combined with a system of local or spot ventilation (see Section 6.3).

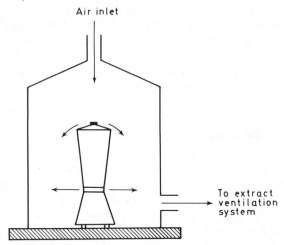

Figure 6.9 A special enclosure to provide ventilation around a blending machine.[72] (Courtesy of Chemistry in Britain)

6.6 Total enclosures (glove-boxes)

The handling of materials in total enclosures is a slow and difficult procedure requiring special skills acquired by long experience. Such enclosures may be required for one or more of the following reasons: (a) to provide high-integrity containment of very dangerous material or of less dangerous material in readily dispersed form such as a powder; (b) to enable a special working atmosphere to be provided in an efficient and economical manner; and (c) to protect the experimental material from contamination by the general atmosphere of the laboratory.

Laboratory Ventilation

Figure 6.10 A high-integrity glove-box (total enclosure) awaiting connection of the compressed-air ejector pump

Total enclosures usually take the form of a virtually leak-tight box with a viewing window[96,99-103] (*Figure 6.10*). The work is carried on through large gauntlet gloves set into glove-ports in the walls of the enclosure. Such enclosures or glove-boxes are usually constructed of mild steel, stainless steel, polyvinylchloride or reinforced plastics. The windows are usually made from plastics, or toughened or laminated glass, and are set in a gasket designed to withstand the pressure difference between the inside and outside of the enclosure.

The gloves, which are the most vulnerable part of the enclosure, need to be made of material which is flexible, leak-tight, resistant to abrasion and resistant to attack by the experimental materials being used. Natural latex rubber has good flexibility, adequate leak-tightness and reasonable resistance to acids, but poor resistance to organic solvents. Neoprene latex rubber has better leak-tightness and a higher resistance to organic solvents and also to ultra-violet light. Milled Neoprene has greater mechanical strength and, for particular applications, butyl rubber and other special materials may be used.

The glove-ports are so designed that replacement gloves can be fitted without loss of the integrity of the containment.

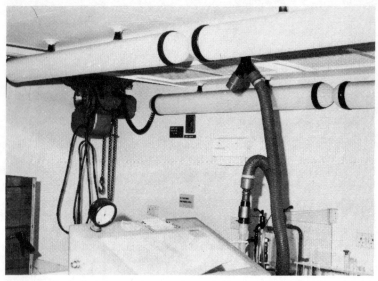

Figure 6.11 A glove-box extract system using a compressed-air ejector pump discharging into the main extract ductwork

There are two principal methods of putting materials into or out of glove-boxes: (1) via posting hatches, (2) by using a sealed-bag technique.

The method chosen depends on a number of factors, including the toxicity and other properties of the materials being transferred. Posting hatches (*Figures 3.8* and *6.10*) provide a rapid means of working but are potential sources of contamination; the sealed-bag technique is generally more reliable.

In an enclosure designed to contain dangerous material, the pressure is maintained at 120–1200 Pa (0.5–5 in w.g.) below atmospheric. Only low rates of air-flow are needed during normal operation to balance the small leakage of air into the enclosure or to balance any gas introduced slowly through a filter or non-return valve. The extracted air or gas must be taken through a filter, sometimes mounted inside the enclosure, and is removed by a compressed-air ejector pump (*Figure 6.11*) or small fan to be discharged into the main fume-extraction system in open-circuit systems. When special atmospheres are needed, a closed-circuit system may be preferable on grounds of economy.

Laboratory Ventilation

In a small glove-box the volume of the gloves themselves is an appreciable fraction of the total; consequently care must be taken not to move a glove into the enclosure too rapidly in case the pressure differential is not maintained by the extract system. In elaborate glove-box facilities, it is possible to arrange for the extract system to automatically increase the rate of air-flow to compensate for a catastrophic failure of containment such as a tear in a glove; the system increases the inflow of air through the hole to about 0.5 m/s (100 ft/min), which should be a sufficiently high velocity to prevent the escape of the dangerous material.[100]

Services to a glove-box must be provided in such a manner that they do not cause air leaks; in extreme cases it may even be necessary to seal the leak-path through electrical cables themselves or to use sealed bulkheads.[100] Controls for all

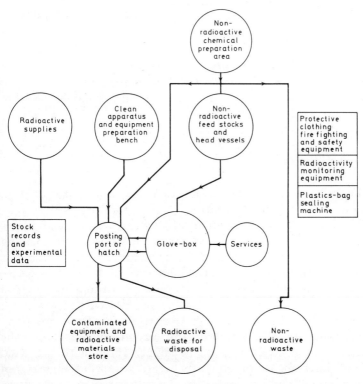

Figure 6.12 A schematic diagram showing the principal ancillary functions and equipment associated with work using a glove-box

Laboratory Ventilation

services should be mounted outside the enclosure. Water supplies can be fitted with safety devices to prevent reverse flow if the supply pressure fails. Other plumbing connections will be needed for head vessels and for feed stocks mounted outside the enclosure. Connection to drain must be indirect, to prevent accidental discharges, and can be through a water ejector pump.

An explosion in a glove-box would have serious consequences; gas burners are therefore not used and, if flammable gases or liquids are introduced into the enclosure, either an inert atmosphere is provided or all sources of ignition are strictly excluded.[104]

A glove-box facility requires a generous allocation of space around the glove-box, not only to give a worker room to withdraw his hands from the gloves but also to accommodate ancillary items (*Figure 6.12*). Also, total enclosures often have glove-ports on more than one side and so access to more than one face is necessary. This means that they are often best placed as free-standing units with services provided from the ceiling.

In microbiological work total enclosures or safety cabinets are used to contain micro-organisms pathogenic to man; the special features required in such applications are discussed in Appendix B.

6.7 Air inlet systems

The air extracted from a laboratory has to be replaced in an appropriate manner. In virtually no laboratory is it permissible to employ a ventilation system which recirculates extracted air. Also, great care must be exercised in locating air inlet points so that fumes discharged from the building or from another source are not collected.[105]

The simplest and cheapest method of introducing air relies on 'natural ventilation' — that is, air allowed to enter in an uncontrolled manner through open windows. There are many drawbacks to this method; the air introduced cannot be filtered or heated; the rate of supply varies with the wind conditions and is consequently erratic; the windows may not be in suitable positions to distribute the fresh air throughout the laboratory or to prevent draughts near fume-cupboards.

Laboratory Ventilation

Some designs take air from an internal corridor through a transfer grille in the wall or the door between the laboratory and the corridor. This is also fairly cheap and the air in the corridor may be warm. However, if the corridor is used by many people, the air taken into the laboratory will be to some degree dirty even if the corridor is supplied with filtered air; furthermore, a simple transfer grille without a fire — and smoke — protection device fitted in it constitutes a weakness in the fire and smoke barriers between the laboratory and the corridor and may not be acceptable to the Fire Brigade.[212] An intumescent honeycomb cell[107] may be inserted in the grille to maintain the integrity of the fire barrier but it does not provide a smoke barrier at normal ambient temperatures. Some attempts have been made to deal with the dirt problem in this type of arrangement; a coarse impact filter has been inserted in the grille, but, because of its resistance, a large proportion of the air will leak into the room through the various cracks around the doors and windows and from the builders' ducts. The air-resistance through the filter can be reduced by substituting an electrostatic precipitator for the impact filter, but such a design is getting rather unwieldy.

In systems which have to be carefully designed because of the cleanliness required, a fan-assisted inlet system is essential; the fresh air is usually drawn from a location fairly high up on the building, but it must be remote from fume discharge points. The air can be filtered, heated and humidified as requried by the application.[7] When one decides on the degree of filtration needed for the inlet air, in addition to any requirements arising within the laboratory itself, the presence of filters in the extract system must also be considered; if absolute filters are fitted in the extract system, a similarly high grade of filter should be provided in the inlet to reduce the frequency with which the potentially contaminated extract filters have to be changed.

A high volume rate of air-flow is also a reason for providing a mechanical inlet system; to avoid draughts which may adversely affect the staff or the work, the incoming air should be distributed in ductwork and introduced through diffusers suitably disposed about the laboratory. The velocity of the air in the space up to 2 m above the floor level should preferably be between 0.1 and 0.2 m/s,[7] and the volume

Laboratory Ventilation

control dampers on the diffusers should be lockable so that the settings are not disturbed when they are cleaned. The velocity of the air through the diffusers should not be so high that it causes a significant increase in the noise level.[49] For very high rates a ventilated ceiling or laminar-flow techniques may be more appropriate (see Chapter 8).

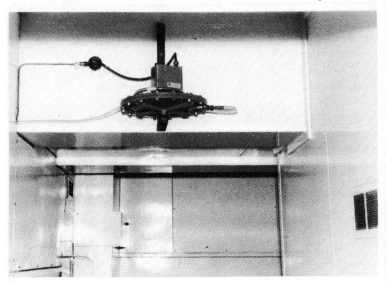

Figure 6.13 A pressure-differential sensing device which maintains the laboratory at a slightly lower pressure than adjacent areas

The extract and inlet systems are generally interlocked — for example, by using an air-flow sensor in the extract (see Section 7.3) — so that an appropriate supply of air is automatically available when the extract system is in use. Provision may be made for the extract system to operate alone if the inlet system fails, but the converse, with the inlet system operating alone, is not desirable, because any contaminants present would be dispersed in an uncontrolled manner.

The extract rate is usually adjusted to be slightly greater than the supply rate so that the pressure in the laboratory is slightly below that in adjacent areas. Also, in a suite of rooms the rates are adjusted so that the areas of highest hazard are at the lowest pressure in order to limit the transfer of contamination to 'cleaner' rooms. In elaborate installations, particularly if the number of fume-cupboards in operation at

Laboratory Ventilation

one time is variable, a pressure-differential sensing device (*Figure 6.13*) and other automatic regulators may be incorporated in the design to maintain the required balance between the extract and inlet systems. Automatic regulation is particularly relevant in systems where dirt-laden filters appreciably alter the resistance and, hence, the rate of air-flow; this point is dealt with in Section 7.2.

As it may be necessary to switch the ventilation system *on or off* following an accident,[108] clearly labelled start and stop switches are often placed near the exit of the laboratory (or in the anteroom if provided) in addition to any switches mounted near the fume-cupboard.

Protective coatings are not required for inlet ductwork to the extent that they are for fume extract ductwork, but streamlined air-flow to reduce the fan power and noise reduction are important. In some cases silencers will be necessary in the supply system, but this is less likely if slow-speed centrifugal fans are installed.

Seven
Fume Extraction and Dispersal

7.1 The extract system

The contaminant having been captured in, say, a fume-cupboard in a laboratory, it is then necessary to convey it safely and without the production of excessive noise[49] through ductwork along the shortest practicable route to a discharge point from which it can be adequately dispersed in an acceptably low concentration. The complexity required in the extract system will depend on the nature of the application, and *Figure 7.1* shows schematically the various elements which might be found in a fume extract system, depending on the intended use. In sequence, they are:

1. The fume-cupboard itself (see Section 6.4).
2. An automatic fire-extinguisher which might be necessary in selected applications.
3. A filter (see Section 7.2).
4. A manometer to monitor the pressure drop across the filter.
5. A damper to automatically compensate for the increase in the resistance of the filter as it accumulates dust.
6. An air-flow sensor (see Section 7.3).
7. An automatic fire-damper (see Section 7.4).
8. A manual setting-up damper with locking screw to ensure that the damper is not moved by vibration. In some instances a locking screw may also be necessary to prevent unauthorised re-setting of the damper.
9. Generously radiused bends and an absence of contortions in order to minimise the resistance of the ductwork (see Section 7.5).

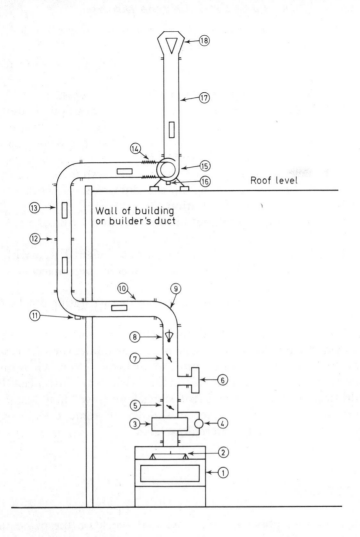

Figure 7.1 A schematic diagram of a fume-cupboard extract system showing all the various elements which might be incorporated, depending on the intended use: 1, fume-cupboard; 2, automatic fire extinguisher; 3, filter; 4, manometer; 5, automatic damper to compensate for changing resistance of filter; 6, air-flow sensor; 7, fire-damper with approved manual override; 8, manual setting-up damper with locking screw; 9, generously radiused bends; 10, minimum of horizontal ductwork; 11, drain connection; 12, ductwork, at negative pressure, with air- and water-tight gaskets at joints; 13, notices stating hazard and if permit to work is required; 14, flexible coupling in ductwork; 15, centrifugal fan; 16, drain connection; 17, tall discharge stack fitted with silencer if necessary; 18, high-velocity discharge nozzle

Fume Extraction and Dispersal

10. Horizontal ductwork kept to the minimum length practicable.
11. A drain connection to take away condensates or rain water.
12. Air- and water-tight joints in the ductwork.
13. Notices stating the nature of the hazardous material which may be in the ductwork and giving instructions about the precautions to be taken or the work-permit to be obtained before maintenance or other work may be carried out on the extract system.
14. A flexible coupling in the ductwork to reduce the transmission of vibration.
15. A centrifugal extract fan (see Section 7.6).
16. A drain connection in the fan casing.
17. A discharge stack of appropriate height fitted with a silencer if the noise level would be unacceptable without one (see Section 7.7).
18. A high-velocity discharge nozzle or rotating cowl (see Section 7.7).

Few systems will need all these elements, and a careful and informed appraisal is necessary of, for example, such items as the fire-damper, the filter and the air-flow sensor, when one specifies the requirements in a particular instance. In addition, extract systems for special purposes may require further items, e.g. a water washdown to remove explosive compounds formed by perchloric acid.[93, 109]

7.2 Filtration

Most fume-cupboard extract systems work on the principle of 'dilute and disperse'. Occasionally, however, particulate matter has to be retained and 'absolute' filters[110, 111] are fitted, preferably near to the fume-cupboard end of the ductwork. Such filters should have appropriate chemical and fire resistance and may be preceded by a pre-filter to extend the life of the main filter. If a high degree of filtration is essential, the filter cell should be tested before use in case of damage during delivery or storage, and particular attention should be paid to the efficacy of the gasket which seals the filter cell to the filter housing.

Fume Extraction and Dispersal

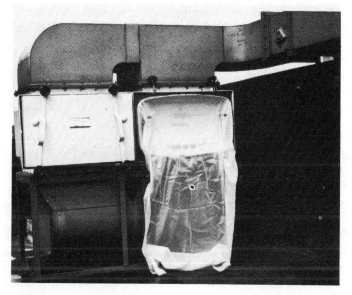

Figure 7.2 Two filter housings of the 'safe-change' type, one showing the PVC bag into which the contaminated filter is withdrawn.[13] *(Courtesy of Koch-Light Laboratories Ltd)*

The filter housing must be well made and of the 'safe-change' type which enables a contaminated filter to be withdrawn into a sealable container such as a PVC bag[13,112] (*Figure 7.2*). The PVC bag can be heat-sealed either by a relatively expensive and heavy radio-frequency welder or by a closely controlled resistance-heated welder.[113] Adequate space must be left in front of and, for some types, also behind the filter housing to enable the filter cells to changed. A manometer should be fitted to measure the pressure drop across the filter so that an indication is obtained of the need to change the filter. In some systems a compensating damper is inserted which automatically balances the increasing resistance of the filter as it clogs up.

7.3 Air-flow sensors

If a fume-cupboard is being used for a particularly hazardous operation, it may be necessary to guard against failure of the extract system or at least to provide a reliable indication that a failure has occurred. An indicator lamp to show that the fan motor is energised or one connected to a centrifugal switch to show that the fan shaft is rotating is not sufficient; the belt drive may have snapped or the fan blades may have disintegrated, and a false indication would be given.

A more reliable indication can be obtained by fitting an air-flow sensor which is then used either to switch in stand-by plant automatically or to activate a warning system. One type of air-flow switch used for this application responds to the difference between the pressure in the extract ductwork and the pressure of the laboratory atmosphere. It contains a mercury tilt-switch mounted on a pivoted vane which is deflected as a result of the pressure difference.[114] The device is essentially an on/off switch and is sensitive down to a pressure difference of 25 Pa (0.1 in w.g.). It is important to realise that this type of sensor responds to *pressure difference* and *not* to the actual air-flow; care must be taken, therefore, to avoid inadvertent obstruction of the ductwork, perhaps by an insecure damper, on the fume-cupboard side of the air-flow switch.

7.4 Fire-dampers

A fire-damper is introduced into ductwork to prevent the spread of hot fumes along the ductwork in the event of a fire.[212, 214] In principle it consists of a damper which closes when the temperature in the ductwork rises sufficiently to melt a fusible link which is restraining the damper. It is advisable to provide a readily operated manual override which enables the damper to be reopened quickly in order, for example, to re-establish extraction to remove fumes after a small fire has been extinguished.

Whether or not to provide a fire-damper in ductwork and, if so, where to position it are subjects on which opinions are divided. On the one hand, it is desirable to reduce the supply of air to a fire and to stop the fire spreading through the duct system. On the other hand, it is important to provide an outlet for the smoke and fumes. Among the factors that have to be taken into consideration are: (a) the material of construction of the ductwork, (b) the route taken by the ductwork and whether or not the ductwork passes from one fire-compartment of the building to another and (c) the

Fume Extraction and Dispersal

presence in the laboratory of expensive or otherwise valuable electronic or other apparatus which would be damaged by corrosive combustion products.

If the ductwork is made of a plastics material such as PVC, it will collapse at an early stage in the fire, and should be protected by a fire-damper. However, if the plastics ductwork is placed within a metal sheath which acts as a fire-barrier, the provision of a fire-damper is less important. In a situation where metal ductwork runs within a fire-protected service duct to an individual fan on an open roof, the balance of arguments would seem to be strongly against providing a fire-damper. However, if the duct is not inside a service duct, but passes through other rooms in an unprotected state, there is more reason to consider fitting a fire-damper.

Where there is long, horizontal ductwork along which hot gases and smoke could be carried to distant parts of the building, the arguments for fitting fire-dampers are stronger than those for vertical ductwork, and the local Fire Brigade is likely to recommend the protection of horizontal ductwork in this way. If the ductwork passes from one fire-compartment to another, a fire-damper will almost certainly be required. It should be noted that in this context the air inlet ductwork may be more important than the extract ductwork. To meet the difficulty of limiting the supply of air to a fire, while at the same time removing smoke and hot gases, a system may be provided which enables the inlet air supply to be cut off, either by means of automatic dampers, including intumescent honeycomb dampers[107] (see Section 6.7), or by the Fire Brigade from a control point, the extract being left operating.

In some situations there are good reasons for it to be preferable to keep the smoke and fumes within the building: for example, in a radioactive laboratory there may be sufficient amounts of a particular radionuclide in a fume-cupboard to be a public health hazard, or at least an embarrassment to the laboratory management, if it were released to the atmosphere. In this instance a fire-damper might well be fitted. The extra cost of decontamination and rehabilitation within the building would be part of the cost penalty of using such materials.

It will be seen from the above notes that the detailed design and use of the building can swing the arguments either way; it is therefore necessary for a detailed discussion to take place among the designers, the users and the local Fire Brigade before the decision to install fire-dampers is taken.

7.5 The ductwork[214]

The locations for fume-cupboards and the routes to be taken by the extract ductwork should be considered during the initial stages of conception of the building ('... services and equipment should more or less decide the shape and design of the building shell itself').[115] Contortions in the ductwork and horizontal sections where condensates may collect and cause corrosion should be avoided unless arrangements are made to drain them. Bends should be generously radiused and any branch entries should be angled to conform with the best engineering practice.[116] It is important to check that the duct-runs which are actually built are the same as those used for the resistance calculations; any last minute re-routing may increase the resistance of the system, thus necessitating fans of higher rating.

The velocity required for the transport of fumes in the ductwork depends on the application and, in particular, on whether or not particular matter is present.[221] High velocities permit of the use of ducts of small cross-section but require larger fans and can result in unacceptably high noise levels.[49] Velocities up to about 7.5 m/s[47] represent a reasonable compromise between noise level and duct size for many applications when gases are being extracted, but velocities two or three times higher are needed to extract particulate matter.[73]

The ductwork should be well made and with the extract fan at the discharge end; the ductwork is then at reduced pressure and consequently any incidental leakage is inwards. This is obviously very important when ductwork is within a building. For ductwork outside a building poorly made components or badly sealed joints may result in the ingress of large quantities of rainwater. It is sometimes advisable to provide the ductwork or the fan-casing with a drain connection; this is especially so if the fan is used intermittently or is not connected into a fume collection duct (see Section 7.7) because rainwater can enter through the discharge nozzle.

Fume Extraction and Dispersal

It is convenient to make metal ductwork in sections about 2 m in length with bolted flanged joints and gaskets of Neoprene or perhaps PVC. Removal of sections for decontamination, disposal or repair is facilitated. Also, in the case of metal ductwork which has a protective coating applied to the inner surface, 2 m is about the maximum length which can be treated in one piece. In the case of asbestos-reinforced thermoplastic materials such as 'Duraform' (see below), patent methods of joining lengths of ductwork have been developed;[117] the Department of Health and Social Security also makes specific recommendations for joining this class of material.[7]

The materials of construction for the extract ductwork should be fire-resistant and of high softening temperature and should have a non-absorbent surface with appropriate resistance to chemical attack. In practice, the choice is inevitably a compromise depending on the application. Masonry and asbestos-based materials have many advantages but may have absorbent surfaces which make them unsuitable unless appropriately treated with a protective coating for applications where contamination is important, as in the case of radioactive substances. Asbestos cement has the advantage of dampening noise because of its weight but it may shatter in a fire.

Plastics materials have good surfaces and good chemical resistance (see *Table 3.3*) and are sometimes selected for these reasons. Unfortunately, they have three serious disadvantages: (1) Some plastics ductwork is not particularly robust, may become brittle through ageing on exposure to the ultra-violet radiation in the atmosphere and is easily fractured by the careless workman. (2) Plastics materials have low softening temperatures − in the case of PVC, about 60°C; this limits the possible applications and also means that the integrity of the ductwork is lost in a fire unless the plastics is supported by or encased in metal. (3) In a fire, plastics decompose with the production of smoke and corrosive products.[84-89, 215]

Some materials − for example, Duraform[118] − have been developed which combine the properties of plastics and asbestos; the plastics gives good chemical resistance and the asbestos maintains the integrity of the ductwork in a fire.

Mild steel ductwork has to be coated to give it appropriate

Table 7.1 SUMMARY OF CHEMICAL RESISTANCE OF PROTECTIVE COATINGS [36,112]

Product	Flash-point before application (°C)	Max. continuous working temperature (°C)	Finish	Resistance to:				
				Inorganic salts and water	Alkalis	Dilute acids	Aromatic solvents	Oils and aliphatic solvents
Durathane [120]	9–29	150	gloss	2	3	1	3	3–4
Epicarb [120]	31	130	semi-gloss	2	1	1	4	4
EpisolveC [120]	51	150	gloss	2	2	2	3	1
Episolve W [120]	46	150	gloss	2	3	3	3	3
Epoxide enamel [120] series 1	22	150	gloss	2	3	3	4	3
Epoxide enamel [120] series 2	22	150	gloss	2	2	1	4	1
Inertol enamel [120]	39	48	semi-gloss	2	1	1	5	4–5
Regnaclor [120]	47	48	satin	2	1	1	5	4–5
Regnavin [120]	45	160	gloss	2	2	2	5	5
Thermoxide [120]	31	180	gloss	2	1	1	3	1
Aromastic [119]	41	93	gloss	2	4	2	4	1
Bituwhite [119]	41	93	gloss	5	6	5	5	4
Comastic [119]	17	71	eggshell	1	1	3	4	1
Epimastic [119]	16	121	dull	1	1	3	4	1
Epoxide enamel [119]	solvent-less	71	gloss	1	1	3	4	2
Hermastic [119]	41	121	synthetic rubber	2	3	2	5	4–5
High-build expoxide [119]	14	71	semi-gloss	2	1	4	1	1

1 = excellent; 2 = very good; 3 = good; 4 = fairly good; 5 = poor; 6 = very poor.

Fume Extraction and Dispersal

chemical resistance. Galvanised steel may be treated with Comastic or shot-blasted black mild steel with Epimastic,[112, 119] which have the properties summarised in *Table 7.1*. In many laboratory installations the amount of air drawn through the extract system provides a large dilution factor; consequently, the corrosive chemicals are in low concentration in the ductwork unless there is a horizontal section or other place where condensates collect.

If the protective coating is carelessly applied to the inner surface of the metal ductwork, it is possible for corrosion of the metal to occur insidiously underneath the coating, with the result that the ductwork may collapse unexpectedly. It is therefore important that ductwork be treated by firms specialising in such work and possessing the necessary equipment for testing the protective coating for flaws after application. It is also advisable to instruct maintenance fitters not to drill into the ductwork after erection.

In a fire the protective coating would be damaged but the metal ductwork ought to maintain its integrity to a reasonably high temperature. The metal provides no thermal insulation, however, and the ductwork could radiate heat to adjacent services in the same builder's duct.

In some applications stainless steel ductwork may be appropriate. For example, stainless steel ductwork rising vertically for about 2 m above the fume-cupboard and fitted with a water washdown has been used in some establishments for work involving perchloric acid.[109] Alternatively, an extract system, including fan, made entirely of PVC[93] can be used to meet the special hazards arising from the use of perchloric acid (see Section 6.4.1).

Metal ductwork installed outside buildings, and consequently subject to atmospheric corrosion, must be adequately coated on the *outer* surface also. This is particularly important at flanges where rainwater may penetrate between the flange and the body of the ductwork.

7.6 The extract fan

In *Figure 7.1* the extract fan is shown at roof level discharging through its own dispersal stack and discharge nozzle; in large installations the extract fan may discharge

Fume Extraction and Dispersal

into a collecting duct and dispersal system as described later (Section 7.7). It is imperative to ensure that the fan rating is adequate to draw the required amount of air through the resistance of the extract system *as actually built*. It is also important to ensure that the fan is sited and mounted so that the noise and vibration generated are not troublesome to the occupants of the laboratory being ventilated or to persons in nearby rooms. Plastics fans (see below) may be quieter in operation than metal ones, and centrifugal fans can be run at low speed to minimise noise production.

If the fan is located in a plant-room on the main roof, the plant-room itself should be ventilated, either naturally or by a suitable fan depending on the particular situation, so that there is no build-up of fumes within the plant-room from slight leaks in the ductwork. If the fan is located in a normally inaccessible position, due provision must be made to permit of safe access for maintenance; this may entail the specification of a ladder and work platform as part of the installation.

The choice of the material of construction for the extract fan is similar to that for the extract ductwork (see Section 7.5); plastics have good chemical resistance but low operating temperatures; metal fans can be treated with a protective coating to limit corrosion. If the fan is mounted outside the building on the roof, the need for it to maintain its integrity in a fire may not be as critical as for ductwork within a building.

The two types of fan which are in common use in extract systems are the axial flow design and the centrifugal design.

Axial flow fans are available as compact units which are easy to install, and some designers have been tempted to place them in totally unsuitable positions immediately above fume-cupboards, with the resultant disadvantages of positive pressure ductwork (with the possibility of leakage of fumes (see Section 7.5) and of noise in the laboratory). The range of ratings available with axial flow fans is limited in the volume of air which can be moved and in the system resistance against which they can work. The maintenance of the fan motor is complicated by the need to check the ductwork for contamination because the motor is in the air stream, unless a bifurcated fan is used.

Centrifugal fans are capable of higher ratings and have the

Fume Extraction and Dispersal

advantage that the motor is outside the airstream. The drive may be by direct coupling or via a belt-and-pulley system to the motor; the latter arrangement has the disadvantage that the belt may snap but the advantage that the extract rate can be adjusted by altering the pulley ratio (provided that the duty remains within the suitable part of the fan characteristic and within the power of the motor).

If the fan is discharging vertically through its individual stack and not into a collecting duct, a pipe connection will be needed to take rainwater from the fan casing to a drain unless the fan is run continuously; in the latter case the efflux velocity, usually in the region of 20 m/s (4000 ft/min), prevents the ingress of raindrops, which have a maximum velocity of about 10 m/s (2000 ft/min).[81,95,105,106]

7.7 Fume dispersal

The advantages gained by capturing a contaminant in a fume-cupboard or other ventilated enclosure are negated if the contaminant is discharged from the extract system in such a manner that it re-enters buildings — for example, through air inlet systems or through open windows. The severity of the precautions taken when fumes are discharged depends on many factors, including the toxicity of the hazardous material, its concentration at release, the nearness of air intakes, the likely pattern of air currents around the building and the possibility, in the case of obnoxious substances, of creating a nuisance rather than a health hazard.

Considerable dilution usually takes place in the extract system from a fume-cupboard; this is especially so if the system is large and incorporates a collecting duct as described below (*Figures 7.4–7.6*). Even so, reasonably effective dispersal is necessary to limit the possible hazard to people and, in the radioactive case, to sensitive counting equipment. Also, it is necessary to prevent the nuisance of objectionable smells at ground and higher levels. With hydrogen sulphide, for example, released at a rate of 100 ml/min at NTP (200 mg/min), 600 m³/h of air is required for dilution to the threshold limit value and 60 000 m³/h (17 m³/s) for dilution to the limit of detection by smell (see Appendix C).

The dispersal system should ensure that the fumes are

taken reasonably clear of buildings and, in particular, away from air intakes, opening windows and other areas to which people have access.[72, 105] The air movements around a building, including the actual distributions of wind speed and direction, would have to be known if a comprehensive design for a fume-dispersal system were required, but information on such very localised conditions is rarely available. A zone of virtually trapped air is formed to the leeward of a building and fumes released into this zone are not carried clear; in consequence, any air supply taken from within the zone is liable to be contaminated by the fumes. Weather conditions obviously affect the dispersal greatly[121] and some general calculations have been made of the concentrations around stacks.[122-124]

Figure 7.3 Fume dispersal stack with discharge nozzle terminating 7 m above the roof of the plant room on top of a five-storey building. The stack is just high enough to release fumes into the airstream passing the building.[13] (Courtesy of Koch-Light Laboratories Ltd)

A rule-of-thumb which is often used is that the height of the discharge point should be one and a half times the height of the building,[125, 126] although some authors consider this too conservative.[81] For a particular building complex, some assessment of the problem can be made by performing smoke tests on site or on a model placed in a wind-tunnel. In practice, a compromise has to be struck between the ideal and such factors as feasible height of stack, probable releases, cost and irate architects and planners! For a building

Fume Extraction and Dispersal

complex six storeys high, for example, with courtyards at ground level between wings of the complex, a discharge stack extending 7 m above the roof level of the plant-room (18 m high) was found to be just sufficient to carry fumes clear of the buildings[72] (*Figure 7.3*). Subsequently it was possible to perform tests on a model of this building complex in a wind-tunnel, where many wind conditions could be simulated;[127] the results suggested that a stack height of 7 m should be just adequate but that 9 m would have been preferable.

Figure 7.4 A duct which collects and dilutes the fumes from 20 fume-cupboards in a wing of the School of Chemistry, the University of Leeds. (Designers: Oscar Faber and Partners)

The collecting duct shown in *Figure 7.4* is connected to a discharge fan (*Figure 7.5*) and terminates in one of the discharge stacks shown in *Figure 7.6*. The stack was designed following the wind-tunnel tests referred to above,[127] and rises about 9 m from the main roof level. It is capable of being extended by another 7 m if subsequently required. The system collects the fumes from about 50 fume-cupboards and discharges 17 m³/s (35 000 ft³/min) with an efflux velocity of 22 m/s (4500 ft/min). One of the sections of the stack incorporates a silencer to reduce the noise level.[128, 129]

Figure 7.5 A discharge fan with a rating of 17 m³/s (35 000 ft³/min). (Designers: Oscar Faber and Partners)

Figure 7.6 Two discharge stacks rising 9 m above the main roof level and capable of being extended a further 7 m.[127] (Designers: Oscar Faber and Partners)

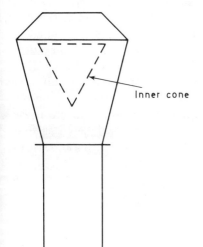

Figure 7.7 A high-velocity discharge nozzle capable of a throw of 20 m in still air[130]

Figure 7.8 Two high-velocity nozzles discharging vertically upwards

Fume Extraction and Dispersal

The design of the nozzle or cowl in which the stack terminates affects the efficiency of dispersal. Horizontal discharges across flat roofs within parapet walls or umbrella cowls which force the fumes downwards are obviously undesirable.[105] A high-velocity nozzle discharging upwards at the top of a stack is preferable (*Figures 7.7* and *7.8*); such nozzles are available in a range of sizes and can be designed to give a long throw even in a moderate wind. In practice, however, one is limited by the high efflux velocity that would be required in high winds to produce a significant throw. This is because of the consequent pressure drop across the nozzle and, hence, the fan power required, the noise generated and the cost. For example, a nozzle such as that shown in *Figure 7.7* would have the following characteristics in still air:[130]

air flow	1.7 m^3/s (3500 ft^3/min)
efflux velocity	15 m/s (3000 ft/min)
throw	20 m
pressure drop	200 pascal (0.8 in w.g.)

A force 7 moderate gale would virtually eliminate the throw.

If the stack is of adequate height, a rotating cowl (*Figure 7.3*) may be of advantage. No vertical throw is produced but, because the cowl rotates with changes of wind direction so that the fumes are released in the direction of the wind, there is a reduction in the power required for the fan motor.

Eight
Laminar Air-flow Clean Rooms and Work Stations

8.1 The need for laminar air-flow

Various requirements for very clean working conditions have given rise to the development of 'laminar air-flow' techniques to provide protection from airborne contaminants for either the work in hand or the worker or, in some cases, both. The term 'laminar air-flow' is defined as an air-flow in which the entire body of air within a confined space moves with uniform velocity along parallel flow-lines.[131] By use of these techniques, a very high degree of control can be exercised over airborne particulate contaminants and, hence, very clean atmospheres can be produced in the laboratory.

Clean rooms, that is laboratories in which the airborne particulate contamination is controlled to a far higher degree than in conventionally air-conditioned spaces, were first used in the electrical engineering industry, where ultra-clean environments were needed for the assembly of small motors, gyroscopes and delicate electronic devices. Subsequently the other great advantage, that of very low bacterial contamination of the air, was utilised in situations where sterility is required; these include work with micro-organisms, the preparation of pharmaceutical products, surgery and the nursing of patients whose immunological resistance to disease has been greatly reduced by chemotherapy or radiotherapy. Further applications are also found in the packaging of foodstuffs, in the handling of toxic chemicals and in the photographic industry.

The early clean rooms were simply provided with filtered air and little control was exercised over the pattern of the air-flow through them. The degree of cleanliness achievable in such a design is limited. *Figure 8.1* illustrates the general pattern of air-flow which develops in such a room between the grilles through which the filtered air enters and the exhaust grilles. Limitations arise because: (a) the turbulent air-flow results in the formation of eddy currents and volumes of virtually static air; (b) the recycling of air in the eddy currents near a source of contamination causes a localised increase in the concentration of airborne contamination; (c) a relatively high differential air pressure is required to prevent increase of contamination from outside the room; (d) only a small amount of self-cleaning occurs; in consequence, the amount of contamination introduced into the room must be limited by rigorous control of the personnel, materials and operations, and an almost continuous cleaning programme is necessary to achieve and maintain levels of contamination which are only moderately good.

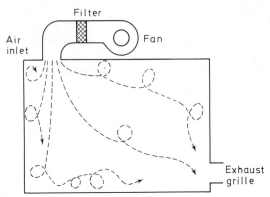

Figure 8.1 A non-laminar air-flow clean room with turbulent air-flow illustrated.[132] *(Courtesy of* Chemistry in Britain*)*

The use of laminar air-flow techniques in clean rooms largely overcomes the above limitations.[133-137] In order to produce a laminar pattern of air-flow, the following principles must be observed: (a) the areas of the air inlet and exhaust grilles must be equal and equal to the cross-sectional area of the space, and (b) the moving air must be contained by walls or sides to the space. These points are illustrated in *Figure 8.2,* where the entire ceiling forms the inlet grille and the perforated floor forms the exhaust grille. The advantages of this arrangement are:

Laminar Air-flow Clean Rooms and Work Stations

1. The parallel flow-lines result in few eddy currents to cause localised recycling of contaminants, thus enabling very low levels of airborne contamination to be achieved and maintained.
2. The air-flow carries any contaminant generated by the work within the room away from the immediate area where the work is being done.
3. The pattern of air-flow precludes cross-contamination between work locations unless one is actually downstream of the other.
4. The self-cleaning action removes airborne contaminants generated within the room or released within it by personnel or from equipment; in consequence, personnel controls may be reduced and the cleaning effort required is lower.

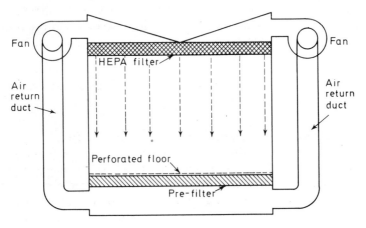

Figure 8.2 *A ceiling-to-floor or downflow laminar air-flow clean room.*[132] (*Courtesy of* Chemistry in Britain)

8.2 Basic design features of laminar air-flow rooms

The large quantity of air required for a laminar air-flow room is introduced into the room through a bank of filters which extends over the entire surface of the ceiling or of one of the walls and which forms the inlet termination of the ventilation system. Great care must be taken with the fitting of the filters; there must be no leakage of air past them and they must extend right up to the walls of the room (*Figure 8.2*), with no solid framework around the periphery to result in unscavenged air near the side walls with the possibility of localised flow in the reverse direction.

Strict laminar air-flow can be obtained only in an empty room; however, if this is achieved, the room should prove satisfactory in use.[136] In a room with an air velocity of between 0.4 and 0.6 m/s (75 and 120 ft/min), the laminar air-flow is soon re-established after passing an obstruction; if air passes along one side only of the obstruction, laminar air-flow is expected to be re-established after a distance equal to six times the diameter of the obstruction, whereas if the air passes along both sides, the required distance is only three times the diameter.

The doors to the room should provide a reasonable seal so that a positive pressure differential of at least 12 Pa (0.05 in w.g.) can be maintained.[131] The fan capacity should be adequate to maintain an outward flow of air when the doors are opened for access. The volume of air necessary to ensure this direction of flow is related to the area of the doorway and to the temperature difference across it, but not to the volume of the room. With a temperature difference of 1°C, a flow of about 15 m^3/min is required for each square metre of door area (50 ft^3/min for each square foot).[138]

The shell of the room should be reasonably free of leaks; it and all furniture should be constructed of easily cleaned materials which do not shed particles or generate static electricity. Anterooms should be provided for the storage of the non-shedding clothing required and for washing facilities.

8.3 Specifications for clean rooms

The performance of clean rooms is usually described in terms of the US Federal Standard No. 209a.[131] Three classes of air cleanliness are prescribed; certain other parameters are specified (e.g. temperature and relative humidity), and others which may need to be specified are listed (e.g. noise, vibration and microbial contamination).

The three classes of air cleanliness are defined in *Table 8.1*. The name of each class derives from the maximum number of particles of size 0.5 μm and larger present per unit volume of the air; hence, the cleanest air is described as 'Class 3.5' in the metric system and as 'Class 100' in the English Imperial system.

The definitions in *Table 8.1* assume a distribution of particle size for each class; these distributions are given in the Federal Standard,[131] but those measured in individual clean rooms may be different. Particle-counting methods are prescribed; light-scattering methods or, for particles of sizes 5.0 μm and larger, the microscopic counting of particles collected on a membrane filter may be employed but the latter is not suitable for monitoring air of cleanliness better than Class 10 000 because so few particles will be present.

Table 8.1 CLASSES OF AIR CLEANLINESS[131]

Class		Maximum number of particles 0.5µm and larger		Maximum number of particles 5.0µm and larger	
English system	Metric system	per cubic foot	per litre	per cubic foot	per litre
100	3.5	100	3.5	*	*
10 000	350	10 000	350	65	2.3
100 000	3 500	100 000	3 500	700	25

*Counts below 10 particles/ft^3 or 0.35 particles/l are unreliable except when a larger number of samples are taken.

The inlet filters should be of the high efficiency particulate arrestance (HEPA) type having an efficiency of at least 99.97% according to the sodium flame test.[139] Pre-filters are provided for the make-up fresh air and for the recirculated air in order to prolong the life of the HEPA filters, which can be as long as 10–20 years.[140] The pre-filters are often located immediately behind the exhaust grilles (e.g. immediately beneath the perforated floor in *Figure 8.2*); this arrangement helps to maintain the room at a positive pressure with respect to its surroundings.

8.4 Types of laminar air-flow clean room and work station

Laminar air-flow is used in various geometrical configurations, depending on the application. The Federal Standard[131] implies that the air velocity should be 0.46± 0.10 m/s (90 ± 20 ft/min) irrespective of the configuration, but other references cite other velocities as noted below.

1. *Vertical downflow clean rooms* In a vertical downflow clean room (*Figure 8.2*) the air enters through the HEPA filters which form the entire ceiling and leaves through the entire area of the perforated floor. The air velocity is usually 0.46 ± 0.10 m/s (90 ± 20 ft/min), which results in unidirectional ventilation at some 0.15 room air-changes/s (540 changes/h) in a room 3 m high. The vertical arrangement is capable of providing a very high degree of air cleanliness, because any contaminant generated at the work place is immediately carried downwards and out of the room.

2. *Horizontal crossflow clean rooms* In this arrangement the HEPA filters form one wall and the air is extracted through the opposite wall. Near the inlet filters, Class 3.5 (Class 100) air cleanliness can exist, but in a room 12–18 m (40–60 ft) long the general level expected is Class 350 (Class 10 000), even with an air velocity of 0.5–0.7 m/s (100–140 ft/min).[136] Care is therefore needed to ensure that the 'dirtiest' work is placed nearest to the exhaust grilles; otherwise contamination of work adjacent or downstream would occur.

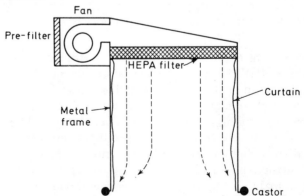

Figure 8.3 A transportable curtained downflow unit.[132] (*Courtesy of* Chemistry in Britain)

3. *Transportable curtained downflow units* A transportable clean-room unit in which Class 3.5 (Class 100) cleanliness can be achieved has been developed from the vertical downflow room by mounting the blower and HEPA filters on a mobile framework draped with plastics curtains to form the walls (*Figure 8.3*). When used within a building, the unit generally

does not interfere with any existing air-conditioning plant, because it merely recirculates the air. When used out of doors or on a dirty floor, the air velocity has to be adjusted to prevent dirt rising to the work areas. The noise and vibration levels of these units may restrict their use.

4. *Horizontal crossflow tunnels* A crossflow tunnel is probably the cheapest clean room to construct. A bank of HEPA filters acts as an end wall for the tunnel, which can be formed out of transparent plastics sheet on a rigid framework. The sides of the tunnel should be sealed to the floor and the far end left open so that the air is exhausted into the surrounding area. With an air velocity of 0.5–0.65 m/s (100–130 ft/min), Class 350 (Class 10 000) cleanliness can be achieved.[136]

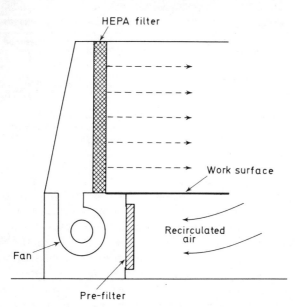

Figure 8.4 *A clean work station with horizontal laminar air-flow*

5. *Clean work stations* A clean work station is a bench or working enclosure having its own filtered air supply (*Figure 8.4*). Thus, a laminar flow of air is sent horizontally (or vertically) over a bench from a bank of HEPA filters by a blower unit which recirculates air drawn from the room (*Figure 8.5*). Class 3.5 (Class 100) conditions can be achieved and several stations can be placed side by side to form a

production line. If the work station is placed in a reasonably leak-free room, the general cleanliness in the room may reach Class 3500 (Class 100 000) because of the large volume of air which passes through the filters.

Figure 8.5 *A laminar air-flow clean work station used in the University of Leeds for work with semiconductors. Air introduced through HEPA filters is blown horizontally over the bench and recirculated via pre-filters beneath the bench.*[132] *(Courtesy of* Chemistry in Britain*)*

6. *Laminar air-flow chemical stations* For the type of work station described in (5) above, the emphasis is on the protection of the work in hand; in some cases it is necessary at the same time to protect the personnel from the work if the latter involves toxic or allergenic materials. In some of the laminar air-flow chemical stations the filtered air is introduced through the ceiling of the cupboard and exhausted through a perforated work surface (*Figure 8.6*); in others the air is directed from one side wall to the opposite wall.[42]

7. *Laminar air-flow microbiological safety cabinets* Several designs have been suggested for laminar air-flow microbiological safety cabinets in which pathogenic organisms are manipulated under sterile conditions.[137] HEPA filters are used in both the supply and extract air systems.[142,143] The systems have to be kept carefully in balance because, on the

Laminar Air-flow Clean Rooms and Work Stations

one hand, an excessive rate of supply can discharge organisms into the laboratory atmosphere and, on the other, an excessive rate of extract can draw unsterilized air into the cabinet.[219] In some designs the work surface protrudes forward and is shaped to introduce a proportion of room air in a direction almost parallel to that of the laminar air-flow (*Figure 8.7*). This unsterilized air, however, should not contaminate the product, because of the absence of lateral mixing in a laminar air-flow, and its introduction ensures that organisms are not discharged into the laboratory if there is a slight deterioration in the performance of the extract system. In others more elaborate arrangements are made in order to enable a solid work surface to be used in place of a perforated one (*Figure 8.8*). Further comments on microbiological safety cabinets are included in Appendix B.

Figure 8.6 A laminar air-flow chemical work station. Air introduced through HEPA filters is discharged vertically downwards and passes through the perforated polypropylene work face.[132,141] *(Courtesy of* Chemistry in Britain *and Microflow Ltd)*

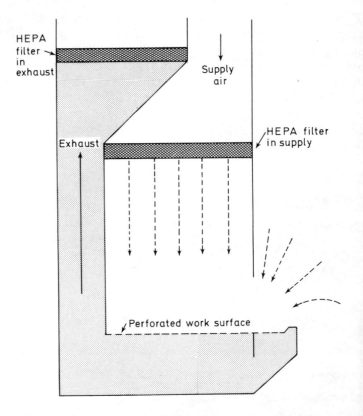

Figure 8.7 A laminar air-flow microbiological safety cabinet with HEPA filters in the supply and in the exhaust air systems. The cabinet provides both sterile working conditions and protection of the worker against the organisms being manipulated[144,145]

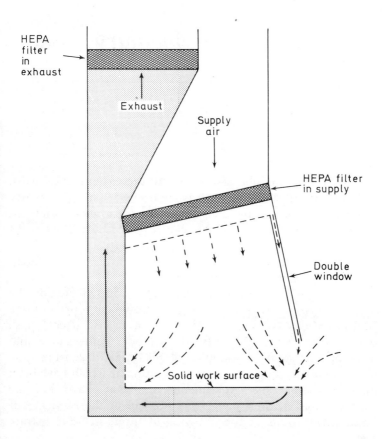

Figure 8.8 A laminar air-flow microbiological safety cabinet with HEPA filters in the supply and in the exhaust air systems and fitted with a solid work surface. The cabinet provides sterile working conditions and protection of the worker against the organisms being manipulated.[132, 145, 146] *(Courtesy of* Chemistry in Britain)

Nine
Stores and Other Ancillary Areas

9.1 Supporting areas

In addition to the laboratories where the experimental work is done, several supporting areas are necessary in a laboratory complex. These will include: (a) stores for a wide variety of items, some requiring special features for safety reasons, of which some are subject to statutory controls, e.g. solvent stores, radioactive materials stores; (b) waste disposal facilities; (c) workshops and plant rooms.

The design of workshops and plant rooms is outside the scope of this book. However, if a hazard which is present in a laboratory can be transmitted to, or otherwise affect, these areas, steps must be taken to protect maintenance and workshop personnel. Examples of this type of hazard include the servicing or repair of items of equipment which have been used in laboratory experiments and may have become contaminated with a dangerous substance, the servicing of fume extract fans or the changing of filters in fume extract ducts.

Where necessary, special provision must be made for the storage of equipment used in hazardous areas by cleaning staff.

The number of stores, the size and layout of these stores and the detailed design of them will depend on the number and type(s) of laboratory to be served. The basic requirements inside the building are: (a) a reasonably central location giving convenient access to all the laboratories in the complex, and (b) proximity to the goods lift in a multistorey building; and, externally, (c) ready and safe access for large vehicles, and (d) adequate safe unloading facilities for

Stores and Other Ancillary Areas

heavy items. In addition there must be good security on both the external and the internal approaches to the store.

Storekeeping involves a substantial amount of record-keeping, and adequate provision must be made for the office work and filing to be carried out conveniently and efficiently. The chief storeman will require to be connected to both the public and any private telephone system in the building.

There are specific requirements for particular types of store; these requirements and those for waste disposal facilities are summarised in the later sections of the chapter.

Wherever hazardous materials are stored, information on their characteristics and details of emergency procedures and first-aid measures must be readily available to the stores personnel. This may entail the provision of notice boards, protective clothing, lockers, safety equipment cupboards and first-aid boxes. An emergency warning system may be required.

Handwashing facilities should be readily available to all stores staff, preferably in the stores.

If it is expected that large and bulky items will be brought into the stores, sufficient space must be provided for suitable trolleys to manoeuvre in the gangways and appropriate lifting tackle must be provided.

Shelves and racks should be laid out in island formation rather than in peninsular formation. This not only allows of better access, but ensures that if an unpleasant chemical is spilled, the storeman can escape by retreating from the contaminated area.

9.2 Solvent stores

There are three types of solvent store, from the point of view of the user: (a) the main solvent store, which will be outside the laboratory building; (b) the transit store, from which the users draw their immediate requirements; and (c) the solvent cupboard, in the laboratory, in which the solvents are kept temporarily.

The main solvent store must cater both for fresh solvents and for waste solvents awaiting disposal. The main store and the transit store will both need to have appropriate dispensing facilities associated with them.

Stores and Other Ancillary Areas

An excise store may, for design purposes, be regarded as an ordinary solvent store, set aside for excisable solvents only.

The legal position is that under Section 2(3) of the Petroleum (Consolidation) Act 1928, 'a local authority may attach to any petroleum spirit licence such conditions as they think expedient, as to the mode of storage, the nature and situation of the premises in which, and the nature of goods with which, petroleum spirit is to be stored ... and generally as to the safe-keeping of petroleum spirit.'[68]

Figure 9.1 *Brick solvent stores built to a Local Authority approved design*

Local Authority conditions for issuing a licence are based on the Model Code of Principles of Construction and Licensing Conditions (Part 1) for the Storage of Cans, Drums and other Receptacles.[147]

Some Local Authorities permit quantities up to 50 Imperial gallons to be stored in approved bins, while others set an upper limit of 25 Imperial gallons.

Main solvent stores (1) Stores, for quantities greater than are permitted to be stored in bins, are usually required to be separate structures (*Figure 9.1*). Exceptions to this rule may

Stores and Other Ancillary Areas

be permitted provided that the store is at ground level, is isolated from the rest of the building by fireproof walls, ceiling and floor and opens only to the open air. Approval may also be given for the store to be sited on the roof of a flat roofed building.

A typical example of a Local Authority approved solvent store includes the following details:

> brick walls 230 mm (9 in) thick
> a concrete roof 125 mm (5 in) thick
> high- and low-level ventilation grills fitted with flame-arrestor wire gauze (28 meshes to the inch)
> a floor free from drains
> a door sill that forms with the floor a tray capable of containing a volume of liquid equal to the store contents (*Figure 9.2*)

Figure 9.2 A solvent store doorway, showing the door cill, to retain spilled liquid, and the mild steel door

a door of 6 mm (¼ in) mild steel, or other approved resistant material, opening outwards and set into a frame with rebates (*Figure 9.2*)

flame-proof light fittings[148,149]

Mechanical ventilation is not normally required, but if it is, flame-proof equipment must be used, in duplicate, with automatic switch-over circuitry and a stand-by power supply. Alternatively, extract may be provided using an intrinsically safe method, e.g. compressed-air ejector with appropriate safeguards against failure of equipment.

Where large drums of solvents are to be stored, the door sill should be provided with a ramp to ease the moving of the drums into the store.

When a licence is granted for this type of store, it is usual for the provision of appropriate fire-fighting equipment and materials to be stipulated,[147] e.g. a 10 lb dry powder extinguisher or a 2 gal foam extinguisher.

The amount of solvent permitted in a store of this type is stipulated on the licence. The upper limit is usually negotiable. The licence will require that a notice such as the following be painted on the door of the store in plain block capitals 50 mm (2 in) high:

PETROLEUM SPIRIT
HIGHLY FLAMMABLE
NO SMOKING

and, if appropriate:

SWITCH OFF ENGINE

In large organisations it is advisable to have multi-compartment stores, or several stores in convenient locations. Where large quantities of methanol and ethanol are used, it may be administratively convenient, for excise control, to have separate stores for these solvents. The advantages of multi-compartment stores, or separate stores, are: (a) if there is no general storekeeper, separate groups or departments are able to control their own solvent supplies; (b) in the event of a fire the amount of potential fuel at immediate risk is limited; (c) clean solvent can be stored separately from waste solvents.

(2) Metal solvent bins are acceptable storage receptacles for limited quantities of solvents. These must be sited out of

Stores and Other Ancillary Areas

doors, in a place approved by the licensing inspector. The bin should be out of the direct sunshine, but if this is not possible, a lining of a thermal insulating material should be fitted. A suitable type of bin is shown in *Figure 9.3*.

Figure 9.3 A mild steel outdoor storage bin for up to 25 gal of petroleum spirit: (top) closed; (bottom) open. Note that the bin is lined with tiles to provide thermal insulation against solar radiation

Stores and Other Ancillary Areas

Transit stores In very large buildings transit stores are permitted. These also must be licensed. They should be smaller than the main solvent store and be used *only* as stores for closed containers. Any decanting of solvents must always be done in a properly designed dispensary (see below) or in a suitably ventilated laboratory.

Solvent cupboards There is always a need to keep glass bottles containing flammable solvents in laboratories where such solvents are used. These bottles should not be kept on open shelves or on benches. Consequently, some form of fire-resistant cupboard is necessary in the laboratory.

It has been shown that a properly designed wooden cupboard will provide excellent protection for solvent containers if a fire breaks out in the laboratory.[151] The cupboard must be constructed of thick timber e.g. 12 mm (½ in) plywood, with 19 mm (¾ in) rebates and with hinges and locks which will not allow the door to warp easily in the intense heat of a fire. ALL the panels must be of thick timber: a thin (3-ply) plywood back will not do!

9.3 Solvent dispensaries

Solvents may be purchased in a wide variety of containers, ranging from glass winchester bottles (approximately 2½ litres capacity), to 50 gal metal drums. Common intermediate sizes are 2 gal, 5 gal and 10 gal metal cans. In pilot plant laboratories the unit of issue from the store may be the metal drum, but normally smaller quantities are required. In many laboratories it is usual to issue solvents in winchester bottles, but, particularly in teaching laboratories, this would be too extravagant, and the solvent must be dispensed into smaller bottles e.g. 500 ml or 1 litre. Consequently, safe dispensing facilities must be provided, either for filling glass bottles from large metal containers or for filling small bottles from winchester bottles.

Dispensing from the large (10 and 50 gal) drums will most probably be done in the main solvent store, with an appropriate pump and siphon. Because of the size and weight of the drums, room must be allowed in the store for their handling.

Dispensing from smaller containers will usually be done in the general store conveniently near to the solvent transit

Stores and Other Ancillary Areas

store. The dispensing area should be a room separate from the rest of the store, with fire-resisting walls and ceiling, and entered through a fire-resisting door. This door should open outwards from the dispensing area.

Good ventilation is a primary requirement to capture, dilute and discharge any flammable vapours released. For many solvent vapours the lower explosion limit is in the range 1—5% by volume in air. It is important that clouds or plumes of flammable vapour should be diluted as quickly as possible and prevented from drifting to a source of ignition. Because of the risk of spills in the dispensing area, there must be no unprotected drain openings through which solvents could enter the drainage system.

The heating system must be intrinsically safe, e.g. hot water radiators. All sources of ignition, such as naked flames (including pilot lights), radiant heating elements, unsealed switches and thermostats and other sources of sparks, must be excluded from the room. Electrical fittings should comply with British Standard Code of Practice CP 1003: Part 1: 1964; Part 3: 1967[52] and BS 1259: 1958.[149] An efficient fume-cupboard should be provided capable of retaining vapours which are heavier than air, i.e. both top and bottom extraction is required (see Section 6.4).

9.4 Waste solvent facilities

A suitable collecting service for waste solvents is essential. In practice, this is the reverse of a stores issue system. One of the requirements of the Deposit of Poisonous Waste Act 1972[68]* is that waste solvents shall be collected in containers which enable the contents to be identified easily. The main solvent store may be used for both fresh and waste material, but if large quantities are involved, it is better to provide two solvent stores.

9.5 Gas cylinder stores

Gas cylinders, being well made, heavy, metal objects, give an impression of solid strength which engenders a false sense of

---

*Superseded by the Control of Pollution Act 1974

Stores and Other Ancillary Areas

security. Properly used and cared for, a gas cylinder is safe: millions are in use every day, and the very few accidents which occur can usually be attributed to human error rather than mechanical failure. However, in a fire a gas cylinder can be a major source of danger.

Gases can be placed in a number of groups: (1) inert gases, (2) gases which support combustion, (3) flammable gases, (4) corrosive gases and (5) toxic gases.

If a cylinder becomes overheated in a fire, one of two events is likely: either the regulator valve will fail and release the contents of the cylinder, or the heat will raise the pressure in the cylinder to bursting point and a serious explosion will occur. If the cylinder valve fails, gases in group (1) will help to retard the fire, those in groups (2) and (3) will intensify the fire, and those in groups (4) and (5) will make the task of fighting the fire very much more difficult and dangerous.

The most likely accident with a gas cylinder is that it will be knocked over or dropped.* To guard against this type of accident, there must be an ample provision of racks for securing gas cylinders in a safe manner both in the stores and where they are to be used. It should never be necessary to leave gas cylinders unsecured. It also follows that there should be an adequate supply of suitable trolleys for moving gas cylinders.

To reduce the risks described, properly designed gas cylinder stores and gas cylinder stations must be provided. A gas cylinder station is where gas cylinders in use are kept, and from where the gases are piped to the point of use.

The size and layout of the building will dictate the design policy to be followed. Ideally, all gas cylinder stations should be located outside the building and the gases piped to the point of use (*Figure 9.4*). If the size and shape of the building rule out this method, the next best location is in a service corridor. In the absence of a service corridor, arrangements for gas cylinder racks must be made in the working areas. It is essential to have sufficient space in the racks to clamp *all* the gas cylinders whether they are full or empty, in use or not.

---

*A cylinder of compressed gas contains a large amount of stored energy. If a cylinder is dropped on to a hard surface and the valve snaps off, the cylinder will behave as a ram-jet missile.

Stores and Other Ancillary Areas

Figure 9.4 (left) A gas cylinder station out of doors from which the gases are piped into the laboratory; (right) the gas cylinder station with its cover locked in position to prevent tampering with the controls

A gas cylinder store should, where possible, be a detached building, of non-combustible materials. If the store is part of a larger building, it should be at ground level, be on an outside wall, with a door leading directly to the open. It is also recommended that it be separated from the rest of the building by walls and floors having at least two hours' fire resistance. There should be good natural ventilation with both high- and low-level ventilators. The store should give protection against corrosion and extremes of both heat and cold. It is advisable not to have roof lights other than north lights, to avoid overheating by direct sunlight.[152] The store must be secure. It is desirable that the path from the cylinder store to the entrance of the laboratory building should provide the staff with protection against rain, snow and ice.

The number of cylinders to be stored will determine the size and complexity of the store. Where very large numbers of cylinders are used, separate compartments can be provided for each of the groups of gases described above, with sub-divisions for full and empty cylinders. Alternatively, in a situation such as that found in a university, it may be preferable to have separate compartments for each user department.

The route from the delivery point or main store to the user

must be easily negotiated by the type of transporter available. These may be trolleys for single cylinders, small manually operated fork-lift trucks with specially designed cylinder cradles, full-size mechanical fork-lift trucks or flat-topped trolleys. In multistorey buildings lifts must be big enough to accommodate the largest approved trolley when loaded.

9.6 Chemical stores

Chemical stores should be designed to a higher standard than dry goods stores. Account must be taken of the possibility of containers leaking or being broken. The floor should be designed to be easily cleaned and to have good chemical resistance. A reasonable level of mechanical ventilation should be provided. In stores where chemicals are transferred from large containers to small containers, or where standard solutions are made up, a dispensing area should be provided, designed to the same standards as a laboratory, a fume-cupboard being provided where necessary.

Acids and caustic alkalis must be kept under safe conditions. In a large store where these chemicals are dispensed from bulk containers, there should be a washhand basin, eye-wash facilities, an emergency shower, safety clothing lockers, first-aid materials and a telephone. Suitable neutralising agents and absorbents, e.g. soda ash kept in bins (and a shovel), are needed to deal with spilled acids. The floor must be designed to facilitate cleaning up and disposal of the neutralised liquid or slurry. Consequently, it should be resistant to strong acids and alkalis, be suitable for hosing down and not be slippery when wet.

Preparation rooms which are often required in teaching institutions are best treated as annexes to the teaching laboratories rather than as part of the main store.

Where the Dangerous Drugs Act 1965[153] applies, security measures must comply with its requirements. Many chemicals which are highly toxic do not come under this Act. However, it is important for reasons of general safety that such chemicals be stored securely. If it is intended to operate the store on an 'honour' system, to reduce administrative work, the design of the store must ensure that the law is observed.

Stores and Other Ancillary Areas

Certain chemicals are listed as explosives under the Explosives Act 1923.[154] The method of storage must meet the requirements of both the local Police Force and Fire Authority who issue the licence for storage, and who will inspect the premises periodically, to ensure that the terms of the licence are being observed. Safes to which only named persons have access are normally required for the storage of explosives.

9.7 Radioactive stores

The secure storage of radioactive substances is a statutory requirement in establishments to which the Radioactive Substances Act 1960[164] or Factory Regulations[76,165] apply (see Appendix A). In addition,[16,108] the sources should be kept so that:

1. Only authorised persons, from within as well as from without the establishment, have access.
2. There is reasonable protection against damage from fire, flooding and vermin.
3. The shielding is sufficient to reduce the dose rate to an individual outside the store to less than 0.75 mrem/h (see Appendix A for references to data on shielding).
4. The shielding of the sources within the store is subdivided so that a source can be quickly identified and removed without displacing the shielding from all the other sources within the store.
5. The store is ventilated if any gas or vapour is liable to be released from the sources either as a radioactive decay product or as the result of radiation-induced chemical decomposition of compounds.
6. Liquids are doubly contained in case there is a breakage of a container and stood in trays to limit the spread of contamination if there is a spillage.

Stores may be small under-bench units (*Figure 9.5*) constructed of brick or concrete slabs, with a rebated door of steel or concrete slabs, and provided internally with trays and lead bricks. Where the number of sources to be kept is great, larger, separate rooms will be appropriate. If no gamma-ray-emitting substance is to be stored, and consequently no heavy shielding is involved, a standard metal under-bench unit with secure lock could form the basis of the store if the fire protection was considered adequate.

Certain radiopharmaceuticals have to be stored at low temperature. Often these products emit beta-radiation, which does not require heavy shielding, and therefore a laboratory refrigerator or deep-freeze can be used provided that it is of the spark-proofed type and lockable.

Figure 9.5 *An under-bench store for radioactive materials, constructed of brick and with a rebated door made from two concrete paving slabs*

Siting is important: for example, radioactive stores must be placed well away from low-background counting equipment and from stocks of photographic film, unless there is considerable intervening shielding. In some instances small stores can be fitted into the space below a fume-cupboard working surface (*Figure 9.5*); this location is particularly convenient for providing a ventilation connection into the ductwork exhausting the fume-cupboard.

In assessing the extent of the total storage space required, allowance must be made for the accumulation of radioactive waste before disposal. Depending on local arrangements, a waste store may need direct vehicular access.

9.8 Strong-rooms

Strong-rooms, for the storage of valuable documents and records, should be built in either the basement or the ground floor of a building and be capable of resisting the severest fire for at least six hours and be strong enough to resist the collapse of the rest of the building and its contents on top of them.[156] The doorway should be protected by a special fire-resisting strong-room door designed for the purpose.

9.9 Waste disposal

Several Acts of Parliament[68]* are relevant to the problems of waste disposal. Most of these are administered by the Local Authority, and it is essential that the appropriate officers be consulted as to the quality and quantity of effluents which will be permissible. The principal acts are: The Public Health Acts (1936, 1937, 1961);[68] the Clean Air Acts (1956, 1968);[68] the Deposit of Poisonous Waste Act (1972);[68] the Radioactive Substances Act (1960),[164] which is administered by the Department of the Environment; and Local Government Acts applicable to particular areas. Provision will be required for the disposal of some or all of the following types of waste: normal domestic and office waste; workshop waste, e.g. scrap materials; bulky laboratory waste only slightly contaminated with chemicals; waste chemicals; waste organic solvents, both flammable and non-flammable; waste oil; biological waste; broken glass; or any of the above contaminated with radionuclides.

The following notes refer to some of the methods of disposing of laboratory waste.

Macerators Some biological wastes can be disposed of via the drainage system after maceration. The macerators used are ordinary commercial items of equipment. Their use is restricted by the heavy demands made on the water supply and the need to avoid disposing of infected material in this manner. The waste pipes must be of adequate size and laid with a good fall to the sewer.

---

*Since the completion of the manuscript the Control of Pollution Act 1974 and the Health and Safety at Work etc. Act 1974 have become law and are relevant to this section.

Incinerators The location of incinerators is governed by the need to avoid fire risks and to provide a discharge flue which complies with the Clean Air regulations.[68] The design of the incinerator required depends on the type of waste to be burnt, and expert advice should be obtained before a choice is made. Incinerators for biological materials, e.g. animal carcasses, must take account of the high water content of the tissues. Disposal of waste organic chemicals by incineration is attractive, but has the following drawbacks:

1. There must be no heavy metal oxide discharged from the flue during the incineration.
2. The furnace must be hot enough to ensure total decomposition of the chemicals.
3. Many solvents are non-flammable and must be decomposed fully, otherwise they will be merely vaporised into the atmosphere. If they are decomposed, the resulting products, e.g. hydrogen chloride, hydrogen bromide, hydrogen fluoride, may be highly corrosive and toxic. It may therefore be necessary to provide a sophisticated fume scrubber in the flue to comply with the requirements of the Local Authority.
4. Well-designed and efficient atomisers are essential for feeding the solvents into the furnace flame.

Some radioactive materials can be incinerated, provided that an appropriate authorisation is obtained under the Radioactive Substances Act 1960,[164] but care must be taken to avoid possible hazards arising from the handling and the disposal of the ash or from the contamination of the furnace lining.[155]

Waste oil Fumeless waste-oil burners can be obtained and may be appropriate in specific cases. In some localities there are commercial firms which purchase used oil for recovery; there is usually, however, a minimum quantity which they are willing to collect, e.g. 200 Imperial gallons.

Broken glass Most chemistry and biology laboratories produce substantial amounts of broken glass. This is highly dangerous to unsuspecting cleaners and refuse disposal staff. Broken glass should be kept separate from all 'soft' waste. To this end, special containers and collection and disposal bins must be provided. Such items as hypodermic syringes and needles can be included in the broken glass waste system.

Stores and Other Ancillary Areas

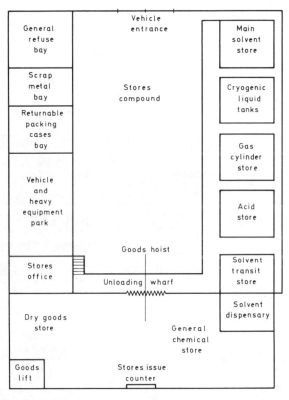

Figure 9.6 *A schematic layout summarising the main stores provisions for a fairly large laboratory complex*

9.10 Stores compounds

A schematic diagram (*Figure 9.6*) summarises the main stores provisions for a fairly large laboratory complex. The detailed layout will to a large extent be governed by local conditions such as the shape of the site and the access roads, and also the management structure of the user organisation. In a very large organisation, or where the laboratories are scattered over a large site, there may be one or more sub-stores.

The main features must include good accessibility for heavy vehicles (*Figure 9.7*). Lorry drivers are often paid piece work rates and are reluctant to deliver to points inconvenient to themselves. There may also be more than one vehicle wishing to deliver materials, and it should not be necessary

Stores and Other Ancillary Areas

for a second vehicle to have to wait for an earlier delivery to be completed before it can gain access to a delivery point. The stores compound should not, for preference, be open to the public, as this introduces security problems.

Figure 9.7 *A store with good vehicle access, ramp and covered loading wharf.*[17] *(Courtesy of Professor S. C. Frazer, the University of Aberdeen, and* Laboratory Practice)

Appendix A
The Requirements for Work with Radioactive Substances

A.1 The nature of the hazard

Radioactive substances emit radiations which have the ability to ionise the materials through which they pass. The ionisation in turn initiates chemical reactions which result in the gross changes observed in the material. If the radiation passes through a living organism or tissue, the chemical reactions and sometimes the initial ionising events themselves cause biological changes which may result in injury or death.

The hazards associated with work involving ionising radiations have been studied extensively since the discovery of X-rays in 1895 and the discovery of radioactivity in 1896. The current recommendations for safe working are contained in a series of reports published by the International Commission on Radiological Protection (ICRP); these recommendations provide much of the basic information required for the national legislation and codes of practice which regulate the use of radiation and radioactive substances in the individual countries.

The recommendations of ICRP[157] introduce the concept of a permissible dose to an individual. This is defined as that dose, accumulated over a long period of time or resulting from a single exposure, which, in the light of present knowledge, carries a negligible probability of severe somatic or genetic injuries; furthermore, it is such a dose that any effects which ensue more frequently are limited to those of a minor nature that would not be considered unacceptable by the exposed individual or by competent medical authorities.

The concept is applied in practice in terms of (a) the maximum permissible doses of radiation which the various parts of the body may receive from sources of radiation external to the body and (b) the maximum permissible body burdens of radioactive substances which the various organs in the body may acquire by inhalation, or by ingestion, or by absorption through the skin (*Figure A.1*). From the values for the maximum permissible body burdens and a knowledge of the metabolic pathways followed in the body by the individual radioactive substances, maximum permissible concentrations of the radioactive substances in air and in drinking-water have been derived.[158-160]

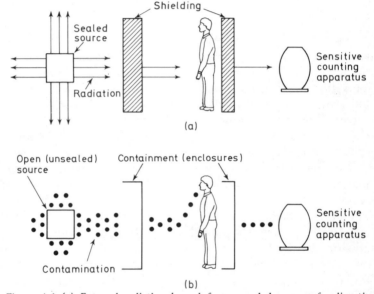

Figure A.1 (a) External radiation hazard from a sealed source of radioactive material; (b) internal hazard caused by contamination from an open (unsealed) source of radioactive material

The design of a laboratory for work with radioactive substances has, therefore, to provide appropriate protection against both external radiation and contamination. The prime concern is to safeguard the health of people in or near the laboratory. In addition, sensitive counting equipment also requires protection; in this case the external radiation dose-rate and the degree of contamination have to be reduced to levels far below those required by considerations of safety.

The hazards which may be met in a radioactive laboratory

The Requirements for Work with Radioactive Substances

arise from the alpha-, beta- and gamma-emissions from spontaneously radioactive substances and from the induced neutron emission produced by certain radioactive substances. (As this book deals primarily with hazardous *materials*, electrical machine sources, e.g. X-ray sets, are not considered.)

Alpha-particles have little penetrating power; consequently, shielding is not usually required to protect persons against the external radiation hazard. Contamination is far more dangerous; when taken into the body, an alpha-particle delivers a large amount of energy to the small volume of tissue through which it travels before it is stopped.

Beta-particles are more penetrating than alpha particles; some light-weight shielding is usually required, and a thickness of 10 or 20 mm of a low atomic number material, such as Perspex or Plexiglass, is often sufficient. The contamination hazard is still important but is less serious than for alpha-particles.

Gamma-ray and neutron emissions are both penetrating (though for different reasons), and relatively massive shielding may be required to reduce the external radiation hazard.

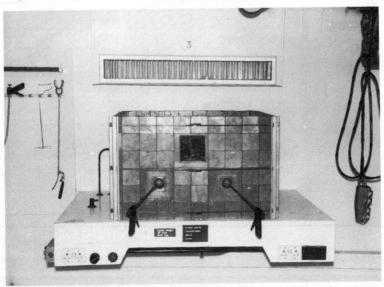

Figure A.2 A shielded enclosure formed from interlocking lead bricks and mounted on a concrete bench. Note the lead-glass window, the remote-handling tongs and the maximum load warning notice

For gamma-radiation the shielding is usually of high-density, high-atomic-number material, e.g. lead, and for neutron radiation it is usually a hydrogenous material, e.g. water or paraffin.

Data on shielding against gamma-radiation are given in a number of publications.[122,161,162] Benches may have to be robust enough to support appreciable weights of such shielding; this requirement may arise in Grade B laboratories (see Section A.3) and will almost certainly arise in Grade A laboratories. The shielding is often in the form of interlocking lead bricks, and an enclosure with external dimensions 1000 mm × 750 mm × 600 mm made of lead bricks 50 mm thick weighs about 1800 kg (4000 lb) (*Figure A.2*). Shielding may also be required as an integral part of the bench-top. This can be achieved by casting the bench-top out of concrete; the concrete, however, must be virtually to the top surface of the bench and finished with only a very thin working surface, in order to avoid a leakage path for radiation between the concrete and any shielding subsequently erected on the bench.

A.2 Legislation and codes of practice

National legislation and codes of practice[163] are based to a large extent on information provided by three sources: (1) the International Commission on Radiological Protection (ICRP); (2) the International Atomic Energy Agency, Vienna; and (3) the specialist organisations within the country concerned (e.g., in the UK, the Medical Research Council, the National Radiological Protection Board and the Atomic Energy Authority).

In the U.K. the principal legislation* affecting the design of premises for work with ionising radiations, excluding that relating to nuclear reactor sites, is:

1. *The Radioactive Substances Act 1960*,[164] administered by the Department of the Environment. The Act requires the registration of premises, with some exceptions such as Crown establishments and hospitals, where radioactive substances are to be kept or used. The conditions imposed include security against unauthorised access to the radioactive sub-

---

*New legislation and codes will arise from the Health and Safety at Work etc. Act 1974.

The Requirements for Work with Radioactive Substances

stances, reasonable protection against fire and the provision of certain standards of finish in the parts of the premises where open (unsealed) sources are to be used. The Act also requires authorisation for the accumulation and for the disposal of solid, liquid or gaseous radioactive waste; consequently, drainage systems and fume discharge arrangements must be approved by the Radiochemical Inspectors.

2. *The Ionising Radiations (unsealed Radioactive Substances) Regulations 1968* No. 780[76] *and the Ionising Radiations (Sealed Sources) Regulations 1969* No. 808.[165] These Regulations, made under the Factories Act 1961,[166] are administered by the Department of Employment; they include specific details of the facilities required by law for work with radioactive substances in factories.

Radiation work carried out in the UK in premises not covered by the Factories Act 1961 is subject to one or more codes of practice.[18,108,167] There are two principal codes (see footnote, p. 124):

1. *The Code of Practice for the Protection of Persons Exposed to Ionising Radiations in Research and Teaching*,[108] which is issued by the Department of Employment and which applies to universities, medical schools, research institutions, technical colleges, Government Departments, the United Kingdom Atomic Authority and industry generally. The Code outlines the general principles of radiation protection and gives guidance on specific types of work, including that involving open sources of radioactive materials.

2. *The Code of Practice for the Protection of Persons Exposed to Ionising Radiations Arising from Medical and Dental Use*,[18] which is issued by the Department of Health and Social Security and the Department of Education and Science. The Code gives guidance on the laboratory and other facilities required in connection with the use of ionising radiations in medical and dental practice and in allied research in hospitals. It also deals with other aspects of radiation protection arrangements in hospitals and with clinical and laboratory techniques.

Advice on the statutory requirements and on the applications of the codes of practice can be obtained in the UK from a number of sources*, including: (a) the Chief Inspector of Factories, (b) the Radiochemical Inspectors of the Department of the Environment, (c) the Advisory and Information Unit of the Department of Employment, and (d) the National Radiological Protection Board.

The use of radioactive materials in schools, establishments of further education and teacher training colleges in the UK is dealt with in Administrative Memorandum 1/65 issued by the Department of Education and Science.[168] Certain regulations also apply.[169-171]

A.3 The grading of radioactive laboratories

In assessing the hazard from open (unsealed) sources and, hence, the extent of the precautions necessary, the following factors are taken into account: (a) the total activity present, (b) the specific activity (i.e. the activity per gram), (c) the radiotoxicity, (d) the chemical toxicity, (e) the chemical form of the radionuclide, (f) the physical form of the radioactive substance, (g) the external dose rate, especially if gamma-radiation is emitted, (h) the competence of the worker, and (i) the type of the operation to be performed.

The types of operation are classified in the following order of increasing potential hazard:

1. Simple storage.
2. Very simple wet operations (e.g. preparation of aliquots from stock solutions).
3. Normal chemical operations (e.g. analysis, simple chemical preparations).
4. Complex wet operations (e.g. multiple operations, or operations with complex apparatus).
5. Simple dry operations (e.g. manipulation of powders) and work with volatile liquids.
6. Dry and dusty operations (e.g. grinding).

Some authorities have introduced a grading system for open source laboratories which takes into account items (a), (c), (f) and (i) listed above.[11,12,18,172] The gradings A, B and C are based on the risks from contamination; they permit

---

*Also from the new Health and Safety Executive.[10]

The Requirements for Work with Radioactive Substances

of work involving up to specified levels of activity for radioactive substances according to the radiotoxicity of the substances and the nature of the experimental work.

Grade A (or Type 1) laboratories are the most elaborate; in a university or similar institution there is perhaps one laboratory available which has been built generally to this standard. They must be very carefully planned and constructed; high standards of design are required throughout — in particular, for the fume extraction and ventilation systems. Otherwise, great difficulties and many frustrations may arise.

Grade B (or Type 2) laboratories are for intermediate levels of activity. Usually several laboratories of this grade are found in a university and a higher proportion in hospitals; they might be used for preparing labelled compounds, for dispensing therapeutic doses of radionuclides or for radiochemistry.

Grade C (or Type 3) laboratories are far more common. They are frequently required in research institutions, and, in many features, differ little from a well-designed and well-constructed laboratory for microchemical or bacteriological work, where cleanliness is also important.

No exact definition of requirement is given in the grading system; it is, however, a convenient basis on which to describe in general terms the standards required, especially during the initial planning stages of a laboratory building. In the earlier chapters of this book, the grade of laboratory envisaged is B/C. There is no sharp dividing line; the higher the activity or the radiotoxicity and the more complex the operations, the more elaborate are the facilities required. Beware, however, the research worker who protests that he is going to use only microcurie amounts for a low-level tracer experiment; he may also need facilities to dispense from very much larger stocks at the beginning of each experiment!

Appendix B
The Requirements for Work with Microbiological Materials

B.1 The nature of the hazard

The accidental spread of infection due to laboratory manipulation of pathogenic micro-organisms was recognised almost from the outset of the scientific study of microbiology in the latter half of the nineteenth century. Occupationally acquired cholera was recorded in 1885,[187, 211] and the first recorded case of blastomycosis (a disease caused by fungus) following accidental self-inoculation was reported in 1903. Since then many thousands of laboratory acquired infections have been recorded in the literature and there have been many deaths.[173-176] Also in recent years there has been concern over the experimental manipulation of the genetic composition of micro-organisms.[217]

Unlike radioactive substances and toxic chemicals, micro-organisms have the ability to multiply. They are very small in size, weighing about 10^{-12} g in the case of bacteria and very much less in the case of viruses. In some instances an infective dose may involve only tens of organisms. Human tissues can be colonised via skin punctures; via the mucosal surfaces of the eyes, nose and throat; and by ingestion or inhalation. In the laboratory it is the last route, inhalation, which assumes the greatest importance; virtually all commonly performed laboratory manipulations give rise to minute splashes and to inhalable aerosols. These are invisible and odourless, and there are no instruments available to give an immediate warning, as, for example, in the case of radioactive contamination.[21, 22, 177-179, 218] In fact, the specific identification of pathogenic organisms is a lengthy process which may necessitate days or even weeks of investigation. In

The Requirements for Work with Microbiological Materials

consequence, a worker may not be aware of the extent of a hazard to himself or to other workers or of cross-contamination until long after an incident, by which time he may himself be infected.

Early methods of contamination control in microbiological laboratories depended largely on the training and skill of the worker to prevent contaminated objects from coming in contact with other materials.[180] However, since it has been realised that airborne contaminants form the major source of hazard, barrier techniques have become all-important.[6,7,15,181,182] The primary barrier usually takes the form of a safety cabinet which may be a total or a partial enclosure, as described in Sections 6.4, 6.6 and 8.4(7), and Section B.3 below. The secondary barrier is provided by the constructional details of the building, including the segregation in the laboratory suite of 'dirty' areas from the 'clean' areas. Both the primary and the secondary barriers have to be provided with suitable ports for the transfer of objects into and out of the controlled area, and, in the case of secondary barriers, people have to pass through too.

B.2 Methods of sterilisation

The most reliable method of producing sterility is by heating, e.g. by autoclaving; however, there are many situations in which it cannot be applied. It is therefore often necessary to resort to less reliable means of sterilisation or to accept partial sterilisation, which is usually referred to as disinfection or decontamination.[183-187]

Agents for the inactivation of micro-organisms can be classified under four headings: (1) heat, (2) vapours and gases, (3) liquid decontaminants and (4) radiation.

1. *Heat* Heat, dry or moist, is considered to be the most effective sterilising agent. The temperatures and times required are generally known, and it should be used whenever possible.[183-186] In the case of contaminated air, for example, the heat may be applied by passing the air through an incinerator.[188,189]

2. *Vapours and gases* Many vapours and gases may be used for sterilisation;[184,186,190-194,220] they are particularly useful in

closed systems, and give good results if used under controlled conditions of temperature and humidity. Care has to be taken because of possible dangers involved; some are explosive (e.g. ethylene oxide), others corrosive to metals (e.g. paracetic acid) and other toxic or carcinogenic (e.g. beta-propiolactone.[191,195]

3. *Liquids* When large areas are involved, decontamination by the mechanical removal of micro-organisms by washing with solutions of disinfectants is often carried out. The germicidal effectiveness of such solutions varies greatly with concentration, temperature, contact time, pH and the presence of organic materials on the areas being treated. In consequence, complete reliance is not usually placed on liquid disinfectants.

4. *Radiation* Ultra-violet radiation, X-rays, gamma-rays and other nuclear radiations are capable of destroying micro-organisms. Ultra-violet radiation of wavelength 254 nm (2540 Å) is effective in certain limited laboratory applications[186,187,196,197] but has very severe shortcomings[187] which must be clearly understood.

X-rays or gamma-rays are usually applied to the sterilisation of packaged goods, e.g. surgical supplies, in special irradiation facilities.

B.3 Microbiological safety cabinets

Microbiological safety cabinets can be considered as specially designed fume-cupboards (Section 6.4) or total enclosures (Section 6.6) or laminar air-flow work stations (Section 8.4(7)). Generally speaking, the microbiologist uses small compact apparatus and carries out manipulations close to the bench surface. As a result, open safety cabinets have been designed which are similar to fume-cupboards but with low fixed apertures through which the air enters.[6,15,17,92,97,98,181,182] As with fume-cupboards, the air velocity must be sufficiently high to overcome the turbulence caused by movements of the operator; in practice, velocities in excess of 1 m/s (200 ft/min) are often necessary because of the high degree of turbulence produced in the relatively small working aperture of a safety cabinet.

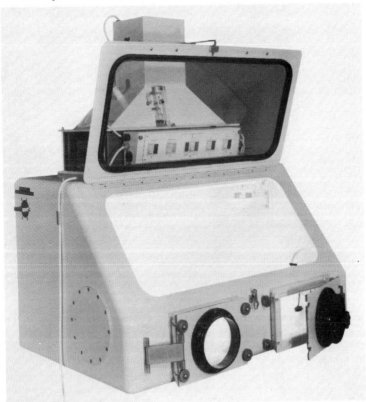

Figure B.1 Microbiological safety cabinet which can be used (a) as an open cabinet with either the glove-ports or the larger glove-port doors open, or (b) as a closed cabinet with gloves sealed into the glove-ports.[182, 198] *(Courtesy of Engineering Developments (Farnborough) Ltd)*

Figures B.1 and *B2* show a microbiological safety cabinet which can be used as an open cabinet with either the glove-ports or the larger glove-port doors open or, alternatively, as a closed cabinet with gloves sealed into the glove-ports.[182,198] With the glove-ports open, the air velocity through the apertures is greater than 0.5 m/s (100 ft/min); with only the gloves removed, the velocity is greater than 1 m/s (200 ft/min). The air-flow through the cabinet is 0.1 m³/s (200 ft³/min) and produces 7 air-changes/min in the cabinet. The air is drawn through pre-filters and then final filters[142,143] with a maximum sodium flame penetration[139] of 0.003%. The fan-motor power rating is 550 W and the fan discharges the air out through the top of the unit. In practice,

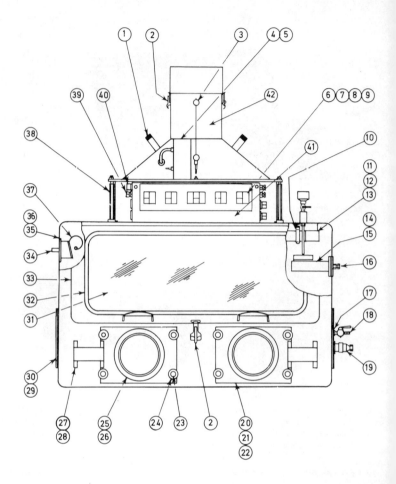

Figure B.2 Components in the microbiological safety cabinet shown in Figure B.1: 1, sample port; 2, toggle catch; 3, retention chain; 4, air-flow gauge; 5, manometer; 6, rocker switch; 7, 8 and 9, neon indicator lights; 10, plastics coated terry clip; 11, 12 and 13, fluorescent tube, end caps and Neoprene sleeve; 14 and 15, formalin vaporiser with heater unit; 16, cable gland; 17, gas tap connector; 18, $\frac{1}{8}$ in BSP gas cock; 19, electric plug and socket set; 20, 21 and 22, glove-port door and seals; 23 and 24, clamping knob and screw; 25 and 26, 6 in glove-port and gasket; 27 and 28, door hinge; 29 and 30, service-port blanking plate and gasket; 31, window; 32, window sealing rubber strip; 33, window frame (hinged); 34, air inlet guard; 35 and 36, flap valve; 37, limit arm; 38, clamp stud; 39, mains input cable gland; 40, 4-pin plug and socket; 41, extract filter; 42, extract fan.[182, 198] *(Courtesy of Engineering Developments (Farnborough) Ltd)*

The Requirements for Work with Microbiological Materials

this type of safety cabinet can be located beneath a fume-extraction hood connected to the main room-ventilation system of the laboratory. Provided that this hood is extracting air at a greater rate than the rate at which the air is being discharged from the cabinet, the room ventilation is unaffected whether the cabinet is operating, and therefore passing a proportion of the air, or not when the room air is extracted directly through the hood.

It is important that the safety cabinet be designed so that the interior, and preferably the filter also, can be sterilised. Ultra-violet radiation (see Section B.2) can be used but is of limited application, especially as it does not penetrate the filter system. Sterilisation is better performed by vapours or gases; for this reason it should be possible to seal open cabinets easily to make them gas-tight. In the above safety cabinet (*Figures B.1* and *B.2*) provision is made to vaporise 25 cm^3 of formalin in a small heater unit built into the equipment.

Closed safety cabinets can be developed into complex gas-tight multi-unit systems with much equipment built into the enclosure.[6,15] Any air drawn into the cabinets passes through ultra-high-efficiency filters and the contaminated air is extracted through similar filters[142, 143, 199] or an air-incinerator.[188,189] Gas-tightness may be tested for by injecting chlorofluorohydrocarbon (Freon) gas into the cabinet to raise the pressure to, say, 1400 Pa (6 in w.g.) above atmospheric and searching for leaks with a halogen gas detector. Less permanent and inexpensive enclosures can be made from plastics sheet supported by frameworks of aluminium tubing.[200]

Another type of microbiological safety cabinet is based on laminar air-flow across the work area (see Chapter 8). Various designs of open cabinet have been proposed; these are summarised in Section 8.4(7) and illustrated in *Figures 8.7* and *8.8*.

B.4 Waste disposal

The sterilisation of liquid waste contaminated with pathogenic micro-organisms is a vital factor before the waste is discharged into the public sewer or even permitted to leave

The Requirements for Work with Microbiological Materials

the laboratory suite. If the micro-organisms are not endemic to the area in which the laboratory is situated, extreme care is needed. Discussion of the details of the various methods of treatment[6,218] is beyond the scope of this appendix. Heating, however, is by far the most reliable and is usually the method chosen in preference to those employing ionising radiation, ultra-violet radiation, ultrasonic radiation or chemicals, including ozone.

Runkle and Phillips describe three different types of system for heat sterilisation:[6] (1) batch sterilisation, (2) continuous high-temperature pasteurisation, (3) constant flow heat-exchanger sterilisation. Each system has its advantages and disadvantages, and expert advice is necessary to establish the most appropriate system for a given laboratory complex.

Appendix C
Threshold Limit Values for Chemically Toxic Materials

An indication of the chemical toxicity of an airborne contaminant is given by the threshold limit value for the substance, which is usually expressed in parts per million by volume in air or in milligrams per cubic metre of air.[201,202] The threshold limit value represents a concentration of the contaminant to which it is believed nearly all workers may be repeatedly exposed, day after day, without adverse effect. Even so, because of the wide variation in the susceptibility of individuals, exposure of a small proportion of individuals at or below the threshold limit value may result in discomfort or injury.

In practice, the threshold limit value is used as a guide to the permissible value for the time-weighted average concentration to which a person may be exposed during the working day, provided that a specific limitation is imposed upon the magnitude of the short-term excursions occurring in excess of the average concentration. There are some substances for which the toxicity is so high that an average value does not provide adequate protection; for these substances 'ceiling' limits or maximum allowable concentrations are specified which must never be exceeded in the air breathed by an individual.

The threshold limit values for a few commonly used chemicals are given in *Table C.1. For comparison, Table C.1* also gives the approximate limits of detection by smell.

Table C.1 EXAMPLES OF THRESHOLD LIMIT VALUES AND LIMITS OF DETECTION BY SMELL [72, 73, 95, 201, 204]

Substance	Threshold limit value (ppm by volume in air)	Approx. limit of detection by smell (ppm)
Ammonia	25	20
Benzene*	25, ceiling limit	100
Carbon monoxide	50	odourless
Carbon tetrachloride*	10	50
Chlorine	1	4
Chloroform	25	–
Formaldehyde	2, ceiling limit	–
Hydrogen sulphide†	10	0.1** (n.b. fatigue)
Methyl bromide	15	††
Sulphur dioxide	5	3

* Marked variation in susceptibility of individuals.
† Very high concentrations result in immediate death.
**Olfactory fatigue after 2–15 min exposure makes it impossible to sense dangerous concentrations by smell.
††The smell is not unpleasant and does not give a definite warning when in dangerous concentrations.

Appendix D
Carcinogenic Substances

Although chemical carcinogenesis has been recognised for 200 years,[205] the legislation and codes of practice have not yet reached the proportions of those which apply to radioactive materials (see Appendix A). In the UK, Factory Regulations[206] identify the more hazardous chemical carcinogens, prohibit the use of some in factories and control the use of others. The use of certain carcinogens is forbidden in schools.[207] No such regulations apply directly to research laboratories (in the UK), where chemists may not always be aware of the potency of the carcinogens with which they work; this problem has been reviewed in the literature.[208] An early code of practice[209] gave some guidance on design requirements.

Appendix E
School Laboratories

The range of materials used in school laboratories is much narrower than in establishments of higher education and in industry. A number of dangerous materials are not permitted in schools by the Department of Education and Science.[168-171, 207, 210, 223] However, this is not to say that all toxic gases and vapours are excluded. Consequently, properly designed fume-cupboards and fume-hoods are essential. One of the authors has actually seen, in a post-1960 school, a fume-hood without any form of air-extract duct or fan! The total absence of effective extract ventilation in school laboratories has earned for Chemistry the school-boy nickname of 'stinks'. This is an unnecessary state of affairs and is a reflection on laboratory designers rather than Chemistry.

Appendix F
Fire Offices' Committee Rules

The Fire Offices' Committee draws up Rules and Recommendations, based on research and testing, which are used as standards by insurance companies. The list of Rules includes the following subjects: policy conditions; automatic fire alarms; construction standards; fireproof doors, lobbies and shutters; drenches; extinguishers; lighting; sprinklers; wired glass and electro-copper glazing.

Further information may be obtained from:

>The Joint Secretaries,
>Fire Offices' Committee,
>Aldermary House,
>Queen Street,
>London EC4

References

1. Munce, J. F., *Laboratory Planning*. Butterworths, London (1962)
2. Nuffield Foundation: Division for Architectural Studies, *The Design of Research Laboratories*. Oxford University Press (1961)
3. Ferguson, W. R., *Practical Laboratory Planning*. Applied Science Publishers, London (1973)
4. Devereux, R. C. de B. and Charlton, R., 'Design of a pharmaceutical laboratory', *Journal of the Institution of Heating and Ventilating Engineers*, 30, 45-59 (1962)
5. Schramm, W., *Chemistry and Biology Laboratories*. Pergamon, London (1965)
6. Runkle, R. S. and Phillips, G. B., *Microbial Contamination Control Facilities*. Van Nostrand Reinhold, New York (1965)
7. Department of Health and Social Security, *Hospital Building Note No. 15: 'Pathology department'*. HMSO, London (1973)
8. *Report on Laboratory Design*. The University of Edinburgh (1970)
9. *An Approach to Laboratory Building*. Laboratories Investigation Unit, Department of Education and Science and University Grants Committee, London (1969)
10. The Health and Safety at Work etc. Act 1974. HMSO, London (1974)
11. Recommendations of the International Commission on Radiological Protection, Report of Committee V on the Handling and Disposal of Radioactive Materials in Hospitals and Medical Research Establishments, 1964. *I.C.R.P. Publication 5*, Pergamon, London (1965)
12. Committee of Vice-Chancellors and Principals of Universities of the United Kingdom, *Radiological Protection in Universities*. The Association of Commonwealth Universities, London (1966)
13. Hughes, D. and Cullingworth, R., *The Design of Laboratories for Radioactive and Other Toxic Substances*. Koch-Light Laboratories Ltd, Colnbrook, Bucks, UK (1971)
14. *Radioisotope Services: A Design Guide*. Department of Health and Social Security, Hospital Building Division, Euston Tower, London, NW1 3DN (October 1973)
15. Phillips, G. B. and Runkle, R. S., 'Laboratory design for microbiological safety', *Applied Microbiology*, 15, 378-389 (1967)
16. Rees, J. M., 'Grading of laboratories and standards of design', in Hughes, D. (Ed.), *Symposium on the Design of Laboratories for Work with Radioactive Substances*, 9-15. Association of University Radiation Protection Officers for British Radiological Protection Association, Birmingham, UK (September, 1970)
17. Frazer, S. C., 'Safety in hospital laboratory design', *Laboratory Practice*, 21, 85-90, 96 (1972)

References

18. *Code of Practice for the Protection of Persons against Ionizing Radiations arising from Medical and Dental Use*, Department of Health and Social Security. HMSO, London (1972)
19. Hughes, D., 'Animal houses for radioactive work', in Hare, R. and O'Donoghue, P. N. (Eds.) *Laboratory Animal Symposia No. 1, The Design and Function of Laboratory Animal Houses*, 119-126. Laboratory Animals Ltd, London (1968)
20. Carr, T. E. F., 'Design of animal houses for radioactive work', in Hughes, loc. cit. (reference 16), 37-43
21. Darlow, H. M., 'Safety in the microbiological laboratory', in Norris, J. R. and Ribbons, D. W. (Eds.), *Methods in Microbiology*, 1, 169-204, Academic Press, London (1969)
22. Collins, C. H., Hartley, E. G. and Pilsworth, R., *The Prevention of Laboratory Acquired Infection*. Public Health Laboratory Service, Monograph Series No. 6, HMSO, London (1974)
23. Malhotra, H. L., *Fire Prevention Science and Technology*, No. 1. Fire Prevention Association, London (March 1972)
24. Hinkley, Patricia M., *The Effect of Coatings on the Ignition Temperature of Fibre Insulating Board by High Temperature Radiation*. Fire Research Note No. 364, Fire Research Station, Borehamwood, UK (1958)
25. Thiéry, P., translated by Goundry, J. H., *Fireproofing*. Elsevier, London (1970)
26. BS 4247: Part 2, *Recommendations for the Assessment of Surface Materials for use in Radioactive Areas: Guide to the Selection of Materials*. British Standards Institution, London (1969)
27. Nairn-Williamson Ltd, Kirkcaldy, Scotland
28. Leeds Fireclay Co. Ltd, Farnley, Leeds 12
29. Hughes, D. and Cullingworth, R., 'Laboratory fittings and waste systems', *Chemistry in Britain*, 8, 470-474 (1972)
30. Hamilton, E. I., 'The design of radiochemical laboratories', *Proceedings of the Society of Analytical Chemistry*, 208-217 (October 1971)
31. BS 4247: Part 1, *Recommendations for the Assessment of Surface Materials for use in Radioactive Areas: Method of Test for Ease of Decontamination*. British Standards Institution, London (1967)
32. Wells, H. and Colclough, W. J., 'Flooring materials for radioactive areas', *UK Atomic Energy Authority Memorandum*. AERE M1065 (1962)
33. Lucas, J. W. and Collins, J. C., 'Selection of surface finishes', in Hughes, loc. cit. (reference 16), 16-22
34. Tompkins, P. C. and Bizzell, O. M., *Industrial and Engineering Chemistry*, 42, No. 8, 1469-1475 (1950)
35. *Durability and Applications of Plastics*, Building Research Station Digest 69. HMSO, London (1972)
36. Hughes, D., 'Design of radionuclide laboratories', *Chemistry in Britain*, 4, 63-66 (1968)
37. Formica Ltd, De La Rue House, 84-86 Regent Street, London W1
38. Imperial Chemical Industries Ltd (Plastics Division), Bessemer Road, Welwyn Garden City, Herts., UK
39. Firth Vickers Stainless Steels Ltd, Staybrite Works, Weedon Street, Sheffield 9, UK
40. British Celanese Ltd, Spondon, Derby, DE2 7BP, UK
41. W. W. Hill, Son and Wallace Ltd, Salford 7, UK
42. Grundy Equipment Ltd, Packet Boat Lane, Cowley Peachey, Uxbridge, Middx, UK
43. Cullingworth, R., *Some Aspects in the Design of Radioactive Laboratories*. Thesis, Institute of Building, London (1970)

References

44. Kitchen waste disposal unit by Permapure (Commercial) Ltd, Stag Lane, Buckhurst Hill, IG9 5TD, Essex, UK
45. Department of Health and Social Security, Hospital Building Division (D3), Room 827, Euston Towers, 286 Euston Road, London NW1 3DN
46. BS 1710, *Specification for the Identification of Pipelines*. British Standards Institution, London (1960)
47. BS 3202, *Recommendations on Laboratory Furniture and Fittings*. British Standards Institution, London (1959), under revision
48. CP 352, *Code of Practice for Mechanical Ventilation and Air Conditioning in Buildings*. British Standards Institution, London (1958)
49. *Code of Practice for Reducing the Exposure of Employed Persons to Noise*, Department of Employment. HMSO, London (1972)
50. Hospital Design Note No. 4: 'Noise control', Department of Health and Social Security. HMSO, London (1966)
51. 'Clenelite' by Crompton-Parkinson Ltd, Crompton House, Aldwych, London WC2
52. CP 1003, *Code of Practice for Electrical Apparatus and Associated Equipment for Use in Explosive Atmospheres of Gas or Vapour other than Mining Applications*:
    Part 1: 1964, *Choice, Installation and Maintenance of Flame-proof and Intrinsically Safe Equipment*
    Part 2: 1966, *Methods of Meeting Explosion Hazard other than by the Use of Flame-proof or Intrinsically Safe Electrical Equipment*
    Part 3: 1967, *Division 2 Areas*
53. *Laboratory Training Manual on the Use of Isotopes and Radiation in Animal Research*. Technical Report Series No. 60, International Atomic Energy Agency, Vienna; HMSO London (1966)
54. *Regulations for the Electrical Equipment of Buildings*. The Institution of Electrical Engineers, London
55. Phillips, R. F., Letter, *Chemistry in Britain*, 3, 89 (1967)
56. The Gas Safety Regulations 1972, S.I. No. 1178. HMSO, London (1972)
57. *A Guide to the Gas Safety Regulations*. British Gas Marketing Division, 326 High Holborn, London WC1V 7PT (1973)
58. *Non-Return Valves for Oxy-Gas Glassworking Burners*. Report No. IM/1/August 1972, British Gas Marketing Division, 326 High Holborn, London WC1V 7PT (1972)
59. van Mourik, J. H. C., 'The formation of nitrous fumes in gas flames', *Annals of Occupational Hygiene*, 10, No. 4, 305-315 (1967)
60. Everton, A. R., *Fire and the Law: An Introductory Guide to the Law relating to Fire*. Butterworths, London (1972)
61. Langdon-Thomas, G. J., *Fire Safety in Buildings: Principles and Practice*. A. and C. Black, London (1972)
62. Marchant, E. W., *A Complete Guide to Fire and Buildings*. Medical and Technical Publishing Co., Lancaster (1972)
63. Lie, T. T., Architectural Science Series: *Fire in Buildings*. Applied Science Publishers, London (1972)
64. Birchall, J. D., *The Classification of Fire Hazards and Extinction Methods*. Benn, London (1961)
65. National Fire Protection Association (USA), *Fire Protection Handbook*, 13th edn. Boston (1969)
66. Home Office (Fire Department),
    (a) *Manual of Firemanship: A Survey of the Science of Fire-Fighting*. HMSO, London.
    Part 1: 'Theory of Fire-Fighting and Equipment'   out of print
    Part 2: 'Appliances'   1973 edn

References

66. (*continued*)

    Part 3:   'Hydraulics and Water Supplies'   1972 edn
    Appendix to Part 3 on Metrication   1971 edn
    Part 4:   'Buildings — Their Construction and Internal Protection'   out of print
    Part 5:   'Communications'   out of print
    Part 6a: 'Practical Firemanship I'   1971 edn
    Part 6b: 'Practical Firemanship II'   1973 edn
    Part 6c: 'Practical Firemanship III'   1970 edn

(b) *The New Manuals of Firemanship*

New manuals are in preparation and will be issued as a series of eighteen books. These are listed below, with the approximate dates of issue.

Book 1: 'Elements of Combustion and Extinction' (Physics of combustion; Chemistry of combustion; Methods of extinguishing fire) — In press

Book 2: 'Fire Brigade Equipment' (Hose; Hose fittings; Ropes and lines, knots, slings; Small gear) — In press

Book 3: 'Fire Extinguishing Equipment' (Hand and stirrup pumps; Portable chemical extinguishers; Foam and foam-making equipment) — 1974

Book 4: 'Pumps and Special Appliances' (Pumping appliances; Practical pump operation; Special appliances) — 1977

Book 5: 'Fire Brigade Ladders' (Extension ladders; Escapes and escape mountings; Turntable ladders; Hydraulic platforms) — 1977

Book 6: 'Breathing Apparatus and Resuscitation' (Breathing apparatus equipment; Operational use of BA; Resuscitation) — 1974

Book 7: 'Hydraulics and Water Supplies' (Hydraulics; Hydrants and water supplies; Water relaying; Appendices) — 1975

Book 8: 'Building Construction and Structural Fire Protection' (Building materials; Elements of structure; Building design) — 1974

Book 9: 'Fire Protection of Buildings' (Fire extinguishing systems; Fire alarm systems; Fire venting systems) — 1974

Book 10: 'Fire Brigade Communications' (Public telephone system and its relationship to the fire service; Mobilising arrangements; Call-out and remote control systems; Radio; Automatic fire alarm signalling systems) — 1974

Book 11: 'Practical Firemanship I' (Practical fire-fighting; Methods of entry into buildings; Control at a fire) — 1976

Book 12: 'Practical Firemanship II' (Methods of rescue; Ventilation at fires; Salvage; After a fire) — 1976

Book 13: 'Fireboats and Ship Fires' (Fireboats and their equipment; Seamanship; Firemanship; Fires in ships) — 1976

Book 14: 'Special Fires I' (Rural areas; Grain, hops, etc.; Animal and vegetable oils; Rubber; Sugar) — 1975

Book 15: 'Special Fires II' (Dusts; Explosives; Fats and waxes; Fibrous materials; Metals; Paints and

References

66. (*continued*)
    varnishes; Plastics; Radioactive materials; Refrigeration plant; Resins and gums) — 1975
    Book 16: 'Special Fires III' (Gasworks; Electricity undertakings; Aircraft) — 1977
    Book 17: 'Special Fires IV' (Fuels; Oil refineries) — 1977
    Book 18: 'Dangerous Substances' (Alphabetical list of dangerous substances) — 1975

67. Fire Prevention Association, Fire Safety Data Sheets.
    FS 6001: 'Portable fire extinguishers: how to choose'.
    FS 6002: 'Portable fire extinguishers: siting care and maintenance'.
    FS 6003: 'Portable fire extinguishers: how to use'.
    FS 6004: 'Fixed fire-extinguishing equipment: the choice of a system'
    FS 6005: 'Automatic fire alarm systems'.
    FS 6006: 'Fixed fire-extinguishing equipment: hose reels'.
    FS 6007: 'Fixed fire-extinguishing equipment: hydrant systems'.
    FS 6011: 'Flammable liquids and gases: explosion hazards'.
    FS 6012: 'Flammable liquids and gases: explosion control'.
    FS 6013: 'Flammable liquids and gases: ventilation'.
    FS 6014: 'Flammable liquids and gases: electrical equipment'.
    FS 6021: 'Explosible dusts: the hazards'.
    FS 6022: 'Explosible dusts: control of explosions'.
    FS 6023: 'Explosible dusts: the elimination of ignition sources'.

Additional Fire Safety Data Sheets are to be issued and further information is available from: Fire Prevention Information and Publications Centre, Fire Protection Association, Aldermary House, Queen Street, London EC4N 1TJ.

68. The Public Health Act 1936
The Public Health (Drainage of Trade Premises) Act 1937
The Public Health Act 1961
The Building Regulations 1972, S.I. No. 317 and amendments
The Building Standards (Scotland) (Consolidation) Regulations 1971, S.I. 2052 (S.218)
The Fire Precautions Act 1971
The Offices, Shops and Railway Premises Act 1963
The Clean Air Act 1956
The Clean Air Act 1968
The Explosives Act 1923
The Petroleum (Consolidation) Act 1928
The Petroleum (Inflammable Liquids and other Dangerous Substances) Order 1947, S.I. No. 1443
The Petroleum (Inflammable Liquids) Order 1968, S.I. No. 570
The Highly Flammable Liquids and Liquefied Petroleum Gases Regulations 1972, S.I. No. 917
Control of Pollution Act 1974
All HMSO, London

69. Publications of the British Standards Institution, London:
    BS EN2: 1972, *Classification of Fires*
    BS 3980: 1966, *Boxes for Foam Inlets and Dry Risers*
    BS 2560: 1954, *Exit Signs for Cinemas, Theatres and Places of Public Entertainment*
    BS 4218: 1967, *Self-luminous exit signs*
    BS 3116, *Automatic Fire Alarm Systems in Buildings:*
        Part 1: 1970, *Heat Sensitive (Point) Detectors*
        Part 4: 1974, *Control and Indicating Equipment*

References

69. (*continued*)
    BS 336: 1965, *Fire Hose Couplings and Ancillary Equipment*
    BS 2599: 1965, *Flax Canvas Unlined Hose for Fire Fighting and Fire Protection*
    BS 3169: 1970, *Rubber Reel Hose for Fire Fighting Purposes*
    BS 3165, *Rubber Suction Hose for Fire Fighting Purposes:*
       Part 1: 1959, *Type A Hose with Partially Embedded Wire*
    BS 476, *Fire Tests on Building Materials and Structures:*
       Part 3: 1958, *External Fire Exposure Roof Tests*
       Part 4: 1970, *Non-combustibility Test for Materials*
       Part 5: 1968, *Ignitability Tests for Materials*
       Part 6: 1968, *Fire Propagation Tests for Materials*
       Part 7: 1971, *Surface Spread of Flame Tests for Materials*
       Part 8: 1972, *Test Methods and Criteria for the Fire Resistance of Elements of Building Construction*
    BS 889: 1965, *Flameproof Electric Lighting Fittings*
    BS 229: 1957, *Flameproof Enclosure of Electrical Apparatus*
    BS 4683, *Electrical Apparatus for Explosive Apparatus:*
       Part 1: 1971, *Classification of Maximum Surface Temperatures*
       Part 2: 1971, *The Construction and Testing of Flameproof Enclosures of Electrical Apparatus*
       Part 3: 1972, *Type of Protection*
    BS 4422, *Glossary of Terms Associated with Fire:*
       Part 1: 1969, *The Phenomenon of Fire*
       Part 2: 1971, *Building Materials and Structures*
       Part 3: 1972, *Means of Escape*
    BS 3251: 1960, *Hydrant Indicator Plates*
    BS 4399: 1969, *Symbols, Dimensions and Layout of Safety Signs for Use in Industry*
    BS 1635: 1970, *Graphical Symbols and Abbreviations for Fire Protection Drawings*

  British Standard Codes of Practice:
    CP 402, *Fire Fighting Installations and Equipment:*
       402.101: 1952, *Hydrant Systems*
       402.201: 1952, *Sprinkler Systems*
       402.Part 3: 1964, *Portable Fire Extinguishers for Buildings and Plant*
    CP 153: Part 4: 1972, *Fire Hazards Associated with Glazing in Buildings*
    CP 3: Chapter 1V: 1948, *Precautions against Fire*
    CP 95: 1970, *Fire Protection for Electronic Data Processing Installations*
    CP 1019: 1972, *Installation and Servicing of Electrical Fire Alarm Systems*
    CP 1003, *Electrical Apparatus and Associated Equipment for Use in Explosive Atmospheres of Gas or Vapour Other than Mining Applications:*
       Part 1: 1964, *Choice, Installation and Maintenance of Flameproof and Instrinsically Safe Equipment*
       Part 2: 1966, *Methods of Meeting Explosion Hazard Other than by the Use of Flameproof or Intrinsically Safe Electrical Equipment*
       Part 3: 1967, *Division 2 Areas* (i.e. areas under very strict control)

References

69. (*continued*)
    Fire Extinguishers:
    BS 138: 1948, *Portable Fire Extinguishers of the Water Type (Soda Acid)*
    BS 740: part 1: 1948, *Portable Fire Extinguishers of the Foam Type (Chemical)*
    BS 740: part 2: 1952, *Portable Fire Extinguishers of the Foam Type (Gas pressure)*
    BS 1382: 1948, *Portable Fire Extinguishers of the Water Type (Gas pressure)*
    BS 3326: 1960, *Portable Carbon Dioxide Fire Extinguishers*
    BS 3465: 1962, *Portable Dry Powder Type Fire Extinguishers*
    BS 3709: 1964, *Portable Fire Extinguishers of the Water Type (Stored pressure)*
    BS 1721: 1968, *Portable Fire Extinguishers of the Halogenated Hydrocarbon Type*
70. Department of Trade and Industry, *Computer Installations: Accommodation and Fire Precautions*, Revised edn HMSO, London (1972)
71. *Report on the Performance of Fire Alarm Systems in Great Britain*, The British Fire Protection Systems Association Ltd, 36 New Broad Street, London EC2M 1NX
72. Hughes, D., 'Laboratory ventilation and fume dispersal', *Chemistry in Britain*, 8, 288-292 (1972)
73. Sax, N. I., *Dangerous Properties of Industrial Materials*, 3rd edn, Reinhold, New York; Chapman and Hall, London (1968)
74. Harvey, B. and Murray, R., *Industrial Health Technology*, Butterworths, London (1958)
75. Hughes, D., 'The design and installation of efficient fume-cupboards', *British Journal of Radiology, Special Report No. 8*, 47, 888-892 (1974)
76. The Ionising Radiations (Unsealed Radioactive Substances) Regulations 1968, S.I. No. 780. HMSO, London (1968)
77. *Report of Working Party on Fume-Cupboard Design*, British Occupational Hygiene Society. In preparation
78. Hewitt, P. J., 'Containment of airborne contaminants by air-flow', *Heating and Ventilating Engineer*, 46, 327-329 (1973)
79. Hewitt, P. J., 'Measurement of face velocity', *The Building Services Engineer* (The Institution of Heating and Ventilating Engineers, London), 40, 83-85 (1972)
80. *Proceedings of the Conference on Laboratory Design for Handling Radioactive Material, November 1951*, BRAB Conference Report No. 3. American Institute of Architects and USAEC, Building Research Advisory Board, National Research Council, National Academy of Sciences, Washington, DC (1952)
81. Horowitz, H., 'Planning for fume-cupboards in the design of science buildings', *Journal of Chemical Education*, 44, A439-A442 (1967)
82. Schulte, H. F., Hyatt, E. C., Jordan, H. S. and Mitchell, R. N., 'Evaluation of laboratory fume-hoods', *Industrial Hygiene Quarterly*, 195-202 (September 1954)
83. McIlroy, R., private communication (1974)
84. 'Corrosion — acid from burning plastics is a new factor in fire costs', *Fire Protection Association Journal*, No. 78, 40-41 (May 1968)
85. 'An international look at corrosion associated with fires involving plastics', *Fire Protection Association Journal*, No. 86, 16-17 (April 1970)
86. 'Plastics — Fire — Corrosion', *Proceedings of the International Symposium and 15th Nordic Fire Protection Day*, The Swedish Fire Protection

References

86. (*continued*)
    Association Skydd 1969. Available from the (UK) Fire Protection Association, Aldermary House, Queen Street, London EC4N ITJ
87. Stark, G. W. V., Evans, W. and Field, P., 'Toxic gases from rigid poly vinyl chloride in fires', *Fire Research Note No. 752*. Fire Research Station, Borehamwood, UK (1969)
88. Woolley, W. D., 'A study and toxic evaluation of the products from the thermal decomposition of PVC in air and nitrogen', *Fire Research Note No. 769*. Fire Research Station, Borehamwood, UK (1969)
89. Woolley, W. D. and Wadley, A. I., 'Studies of phosgene production during the thermal decomposition of PVC in air', *Fire Research Note No. 776*. Fire Research Station, Borehamwood, UK (undated)
90. Schumacher, J. C., *Perchlorates, Their Properties, Manufacture and Uses.* American Chemical Society Monograph Series No. 146, Reinhold, New York (1960)
91. Everett, K. and Graf, F. A., 'Handling perchloric acid and perchlorates', In Steere, loc. cit. (reference 95), 265-276
92. Holliday, Fielding and Hocking Ltd, Leeds LS9 8HA, to the design of the University of Leeds
93. Aire-Movement Co., Queens Square Chambers, Woodhouse Lane, Leeds 2, to the design of the University of Leeds
94. Silverman, L. and First, M. W., 'Portable laboratory scrubber unit for perchloric acid', *Industrial Hygiene Journal*, 463-471 (Nov.-Dec. 1962)
95. Steere, N. V., *Handbook of Laboratory Safety*, 2nd edn. The Chemical Rubber Co., Cleveland, Ohio (1971)
96. *International Symposium on the Radiological Protection of the Worker by the Design and Control of his Environment.* Society for Radiological Protection, Bournemouth, UK (1966)
97. Williams, R. E. O. and Lidwell, O. M., 'A protective cabinet for handling infective material in the laboratory', *Journal of Chemical Pathology*, **10**, 400 (1957)
98. Rees, R. J. W., 'Precautions against tubercular infections in the animal house', *Collected Papers*, **10**, 51 (1961). Laboratory Animals Centre, Medical Research Council, Carshalton, UK
99. Walton, G. N. (Ed.), *Glove-Boxes and Shielded Cells for Handling Radioactive Materials:* A Record of the Proceedings of the Symposium on Glove-Box Design and Operation held at AERE, Harwell, Feb. 1957. Butterworths, London (1958)
100. *Manual on Safety Aspects of the Design and Equipment of Hot Laboratories*, Safety Series No. 30. International Atomic Energy Agency, Vienna; HMSO, London (1969)
101. Barton, C. J., *A Review of Glove-Box Construction and Experimentation.* USAEC Report, ORNL-3070, uc-4-Chemistry (1961)
102. Garden, N. B. (Ed.), *Glove-Boxes and Containment Enclosures.* US Report TID-16020 (1962)
103. *Code of Practice for Unshielded Glove-Boxes*, AE CP59. UKAEA
104. May, F. J., *Solvent Fires and Explosions in Glove-Boxes: Investigation to Determine Minimum Hazardous Quantity.* AERE-R5514; HMSO, London (1973)
105. Clark, J. H., 'The design and location of building inlets and outlets to minimize wind effect and building re-entry of exhaust fumes', *Industrial Hygiene Journal*, 242-248 (May-June 1965)
106. Laws, J. O. and Parsons, D. H., 'Relation of rain-drop size to intensity', *Transactions of the American Geophysical Union*, Part II, 452-460 (1943)
107. *Honeycomb Fire Dampers*, Building Research Establishment Digest 158. HMSO, London (1973)

References

108. *Code of Practice for the Protection of Persons against Ionising Radiations in Research and Teaching*, Department of Employment. HMSO, London (1968)
109. Division of Environmental Health and Safety, University Health Service, University of Minnesota, USA
110. Linder, P., *Air Filters for Use at Nuclear Facilities*, Technical Report Series No. 122. International Atomic Energy Agency, Vienna; HMSO, London (1970)
111. Davies, C. N., *Air Filtration*. Academi Press, London and New York (1973)
112. Hughes, D., 'The ventilation of radioactive laboratories', *Journals of the Institution of Heating and Ventilating Engineers*, 37, 35-39 (1969)
113. Star Engineering (Gosport) Ltd, Star Yard, High Street, Gosport, Hants., UK
114. Delta Controls Ltd, Deltrol Works, 145 London Road, Kingston-on-Thames, UK
115. Zetterberg, H., 'Hospital air-conditioning in Sweden', *Journal of the Institution of Heating and Ventilating Engineers*, 36, 155-163 (1968)
116. *Engineers' Guide*. The Institution of Heating and Ventilating Engineers, London (1970); Book C, Table C4-38 and Table C4-39
117. Henry Hargreaves and Sons Ltd, Lord Street, Bury, Lancs., UK
118. Turner Brothers Asbestos Co. Ltd, P.O. Box No. 40, Rochdale, Lancs., UK
119. Wailes Dove Bitumastic Ltd, Hebburn, County Durham, UK
120. Storry Smithson Group, Bankside, Kingston-upon-Hull, UK
121. Pasquill, F., 'The estimation of the dispersal of windborne material', *Meteorological Magazine*, 90, 33 (1961)
122. *Handbook of Radiological Protection, Part I; Data*, Departments of Employment and of Health and Social Security. HMSO, London (1971)
123. Bryant, P. M., 'Methods of estimation of the dispersal of windborne material and data to assist in their application', UKAEA Report AHSB(RB)R 42 and Addendum. HMSO, London (1964)
124. Brook, A. J., 'The effect of deposition on the concentration of windborne material', UKAEA Report AHSB(S)R 157. HMSO, London (1968)
125. Halitsky, J., 'Gas diffusion near buildings', *American Society of Heating, Refrigerating and Air-Conditioning Engineers Transactions*, 69, 446-485 (1963)
126. Wise, A. F. E., Sexton, D. E. and Lillywhite, M. S. T., 'Studies of air-flow round buildings', Building Research Current Papers, Design Series No. 38, Building Research Station. HMSO, London (1965)
127. Everett, K., 'Fume extraction and dispersal problems', in Hughes, loc. cit. (reference 16), 31-36
128. Sound Attenuators Ltd, Eastgates, Colchester, Essex, CO1 2TW, UK
129. BS 4142, *Method of Rating Industrial Noise Affecting Mixed Residential and Industrial Areas*. British Standards Institution, London (1967)
130. Norvenco Ltd, Tundra Way, Chain Bridge Road, Blaydon-on-Tyne, UK
131. U.S. Federal Standard No. 209a, 'Clean Room and Work Station Requirements, Controlled Environment'. General Services Administration, Specifications Activity, Printed Materials Supply Division, Building 197, Naval Weapons Plant, Washington, DC, 20407 (1966)
132. Hughes, D., 'Clean Rooms', *Chemistry in Britain*, 10, 84-87 (1974)
133. Whitfield, W. J., *Technical Report SC-4673(RR)*, Sandia Corporation, Albuquerque, New Mexico (1962)
134. Whitfield, W. J., *Technical Report SCR-652*, Sandia Corporation, Albuquerque, New Mexico (1963)
135. Whitfield, W. J., *Technical Report SCR-66-956*, Sandia Corporation, Albuquerque, New Mexico (1966)

References

136. McDade, J. J., Phillips, G. B., Sivinski, H. D. and Whitfield, W. J., 'Principles and applications of laminar flow devices', in Norris, J. R. and Ribbons, D. W. (Eds), *Methods in Microbiology*, Vol. I, 137-168. Academic Press, London (1969)
137. Phillips, G. B. and Runkle, R. S., *Biomedical Applications of Laminar Air-Flow*. Chemical Rubber Co., Cleveland, Ohio, USA (1973)
138. Whyte, W. and Shaw, B. H., Building Services Research Unit, The University of Glasgow, Scotland. 'Airflow through doorways', Fourth International Conference on Aerobiology, 1970 (to be published)
139. BS 3928, *Method for Sodium Flame Test for Air Filters*. British Standards Institution, London (1969)
140. Firman, J. E., 'Laminar air-flow in operating theatres', *British Hospital Journal and Social Service Review* (September 1972)
141. Microflow Ltd, Fleet Mill, Minley Road, Fleet, Hants., UK
142. Darlow, H.M., 'Filtration of biological particles from air', *Filtration and Separation*, 3, 303-307 (1966)
143. Decker, H. M., Buchanan, L. M., Hall, L. B. and Goddard, K. R., 'Air filtration of microbial particles', *Public Health Service Publication No. 953*, US Government Printing Office, Washington, DC (1962)
144. McDade, J. J., Sabel, F. L., Akers, R. L. and Walker, R. J., 'Microbiological studies on the performance of a laminar airflow biological cabinet', *Applied Microbiology*, 16, 1086-1092 (1968)
145. Akers, R. L., Walker, R. J., Sabel, F. L. and McDade, J. J., 'Development of a laminar airflow biological cabinet', *American Industrial Hygiene Association Journal*, 30, 177-185 (1969)
146. Coriell, L. L. and McGarrity, G. J., 'Biohazard hood to prevent infection during microbiological procedures', *Applied Microbiology*, 16, 1894-1900 (1968)
147. The Petroleum (Consolidation) Act 1928 and Model Code of Principles of Construction and Licensing Conditions (Part I) for the Storage of Cans, Drums and Other Receptacles. HMSO, London (1971)
148. BS 229, *Flameproof Enclosure of Electrical Apparatus*. British Standards Institution, London (1957)
149. BS 1259, *Instrinsically Safe Electrical Apparatus and Circuits*. British Standards Institution, London (1958)
150. *Electrical Safety Code*. Institute of Petroleum, London (1965)
151. Stark, G. W. V., White, R. W. and Moseley, G. E., 'Wooden laboratory cupboards for the fire protection of solvents', *Chemistry and Industry*, 1173-1174 (16 October 1971)
152. *Fire Precautions in the Storage, Handling and Use of Gas Cylinders*. Technical Information Sheet 3017, Fire Protection Association, London (1966)
153. The Dangerous Drugs Act 1965. HMSO, London
154. The Explosives Act 1923. HMSO, London
155. Bush, D. and Hundal, R. S., 'The fate of radioactive materials burnt in an institutional incinerator', *Health Physics*, 24, 564-568 (1973)
156. *Protection of Records and Documents against Fire*. Technical Information Sheet 1002, Fire Protection Association, London (undated)
157. Recommendations of the International Commission on Radiological Protection, 1965, *ICRP Publication 9*. Pergamon, Oxford (1966)
158. Recommendations of the International Commission on Radiological Protection, Report of Committee II on Permissible Dose for Internal Radiation, 1959, *ICRP Publication 2*. Pergamon, Oxford (1960)
159. Recommendations of the International Commission on Radiological Protection, 1962, *ICRP Publication 6*, Pergamon, Oxford (1964)

References

160. *Evaluation of Radiation Does to Body Tissues from Internal Contamination Due to Occupational Exposure, ICRP Publication 10.* Pergamon, Oxford (1968)
161. BS 4094: Part I, *Data on Shielding from Ionising Radiation.* British Standards Institution, London (1966)
162. *The Radiochemical Centre Manual.* The Radiochemical Centre, Amersham, Bucks.; HMSO, London (1966)
163. World Health Organization, *Protection against Ionising Radiations: A Survey of Current World Legislation.* Geneva (1972)
164. The Radioactive Substances Act 1960. HMSO, London
165. The Ionising Radiations (Sealed Sources) Regulations 1969, S.I. No. 808. HMSO, London
166. The Factories Act 1961. HMSO, London
167. *Radiation Safety in Veterinary Practice: Code of Practice for the Protection of Persons exposed to Ionising Radiation from Veterinary Uses.* Ministry of Agriculture, Fisheries and Food, London (1970)
168. *Memorandum on the use of Ionising Radiations in Schools, Establishments of Further Education and Teacher Training Colleges.* Administrative Memorandum 1/65, Department of Education and Science, London (1965)
169. The Direct Grant Schools Amending Regulations 1965, S.I. No. 1. HMSO, London
170. The Further Education Regulations 1969, S.I. No. 403. HMSO, London
171. The Schools Amending Regulations 1965, S.I. No. 3. HMSO, London
172. *Safe Handling of Radioistopes,* Safety Series No. 1. International Atomic Energy Agency, Vienna; HMSO, London (1962)
173. Sulkin, S. E., 'Laboratory acquired infections', *Bacteriological Review,* 25, 203-209 (1961)
174. Uldall, A., 'Occupational risks in danish clinical chemical laboratories I', *Scandinavian Journal of Clinical Laboratory Investigation,* 33, 21-25 (1974)
175. Skinhöj, P., 'Occupational risks in danish clinical chemical laboratories II: infections', *Scandinavian Journal of Clinical Laboratory Investigation,* 33, 27-29 (1974)
176. *Report of the Committee of Inquiry into the Smallpox Outbreak in London in March and April, 1973.* Command 5626, HMSO, London (1974)
177. Phillips, G. B., 'Prevention of laboratory-acquired infections', in Steere (reference 95) 610-617
178. Chatigny, M. A., 'Protection against infection in the microbiological laboratory'. In Umbreit, W. E. (Ed.), *Advances in Applied Microbiology,* Vol. 3, 131-192. Academic Press, New York (1961)
179. Wedum, A. G., 'Control of laboratory airborne infection', *Bacteriological Review,* 25, 210-216 (1961)
180. Bullock, W., *The History of Bacteriology.* University Press, Oxford (1938)
181. *The P.H.L.S. Protective Cabinet: Notes on Its Installation and Use.* UK Public Health Laboratory Service, London (June 1970)
182. Darlow, H. M., 'The design of microbiological safety cabinets', *Chemistry and Industry,* 1914-1916 (1967)
183. Perkins, J. J., *Principles and Methods of Sterilisation.* Charles C. Thomas, Springfield, Ill. (1956)
184. Reddish, G. F. (Ed.), *Antiseptics, Disinfectants, Fungicides, and Chemical and Physical Sterilisation.* Lea and Febiger, Philadelphia (1957)
185. Sykes, G., *Disinfection and Sterilisation.* Spon, London (1965)
186. Rubbo, S. D. and Gardner, J. F., *A Review Sterilisation and Disinfection as applied to Medical, Industrial and Laboratory Practice.* Lloyd-Luke, London (1965)

References

187. *The Use of Chemical Disinfectants in Hospitals*, Public Health Laboratory Service, Monograph Series No. 2. HMSO, London.
188. Decker, H. M., Citek, F. J., Harstad, J. B., Gross, N. H. and Piper, F. J., 'Time-temperature studies of spore penetration through an electric air sterilizer', *Applied Microbiology*, 2, 33-36 (1954)
189. Harris, G. J., Gremillion, G. G. and Towson, P. H., 'Test new electric incinerator design for sterilising laboratory air', *Heating, Piping and Air Conditioning*, 36, 94-95 (1964)
190. Schley, D. G., Hoffman, R. K. and Phillips, C. R., 'Simple improvised chambers for gas sterilisation with ethylene oxide', *Applied Microbiology*, 8, 15-19 (1960)
191. Bruch, C. W., 'Decontamination of enclosed spaces with beta propiolactone vapor', *American Journal of Hygiene*, 73, 109 (1971)
192. Darlow, H. M., 'Safety in the animal house', *Laboratory Animals*, 1, 35-42 (1967)
193. Perkins, F. T., Darlow, H. M. and Short, D. J., 'Further experience with Tego as a disinfectant in the animal house', *Journal of the Institute of Animal Technicians*, 18, 83-92 (1967)
194. Spinner, D. R. and Hoffman, R. K., 'Method for disinfecting large enclosures with B-propiolactone vapor', *Applied Microbiology*, 8, 152-155 (1960)
195. Dickens, F. and Jones, H. E. H., 'Carcinogenic activity of a series of reactive lactones and related substances', *British Journal of Cancer*, 15, 85-100 (1961)
196. Morris, E. J., 'The practical use of ultra-violet radiation for disinfection purposes', *Medical Laboratory Technology*, 29, 41-47 (1972)
197. Wedum, A. G., Hanel, E. and Phillips, G. B., 'Ultraviolet sterilisation in microbiological laboratories', *Public Health Report*, 71, 331-336 (1956)
198. The Porton-E.D.L. Microbiological Safety Cabinet, Engineering Developments (Farnborough) Ltd, Eelmoor Road, Farnborough, Hants., GU14 7NW, UK
199. Harstad, J. B., Decker, H. M., Buchanan, L. M. and Miller, M. E., 'Air filtration of submicron virus aerosols', *Proceedings of the Sixth Annual Technical Meeting*. American Association for Contamination Control, Washington, DC (May 1967)
200. Phillips, G. B., Novak, F. E. and Alg, R. L., 'Portable inexpensive plastic safety hood for bacteriologists', *Applied Microbiology*, 3, 216-217 (1955)
201. *Threshold Limit Values for 1972*, Technical Data Note 2/72, Department of Employment. HMSO, London (1972)
202. *Threshold Limit Values*. American Conference of Governmental Industrial Hygienists, P.O. Box 1937, Cincinnati, Ohio 45201
203. Browning, E., *Toxicity and Metabolism of Industrial Solvents*. Elsevier, Amsterdam (1965)
204. *Matheson Gas Data Book 1966 and Supplements*. The Matheson Co. Inc., P.O. Box 85, East Rutherford, New Jersey 07073, USA
205. Potts, P., *Chirurgical Observations*. London (1775)
206. The Carcinogenic Substances Regulations 1967, S.I. No. 879. HMSO, London (1967)
207. *Avoidance of Carcinogenic Amines in Schools and Other Educational Establishments*. Administrative Memorandum 3/70, Department of Education and Science, London (1970)
208. Searle, C. E., 'Chemical carcinogens and their significance for chemists', *Chemistry in Britain*, 6, 5-10 (1970)
209. *Precautions for Laboratory Workers who handle Carcinogenic Aromatic Amines*. Chester Beatty Research Institute, Institute of Cancer Research, Royal Cancer Hospital, London (1966)

References

210. Everett, K. and Jenkins, E. W., *A Safety Handbook for Science Teachers*. Murray, London (1973)
211. Koch, R., Second Conference on Cholera, Berlin, 4-8 May 1885. Paper in *Recent Essays by Various Authors on Bacteria in Relation to Disease*, selected and edited by W. Watson Cheyne, New Sydenham Society, London (1886) (Translation from Berliner Klinischer Wochenschrift, No. 37A, 1885)
212. Recommendations relating the design of air-handling systems to fire and smoke control in buildings. *Technical Memoranda 1*, Institution of Heating and Ventilating Engineers (1974)
213. Notes on legislation relating to fire and services in building. *Technical Memoranda 2*, Institution of Heating and Ventilating Engineers (1974)
214. CP 413, *Code of Practive for Ducts for Building Services*. British Standards institution, London (1973)
215. Bowes, P. C., 'Smoke and toxicity hazards of plastics in fire', *Annals of Occupational Hygiene*, 17, 143-157 (1974)
216. *Manual on Radiation Protection in Hospitals and General Practice, Vol. 1: Basic Protection Requirements*. World Health Organization, Geneva (1974)
217. *Report of the Working Party on the Experimental Manipulation of the Genetic Composition of Micro-organisms*. Command 5880, HMSO, London (1975)
218. Harris-Smith, R. and Evans, C. G. T., 'Bio-engineering and protection during hazardous microbiological processes', *Biotechnology and Bio-engineering Symposium No. 4*, 837-855, Wiley, Chichester (1974)
219. Newsom, S. W. B., 'A test system for the biological safety cabinet', *Journal of Clinical Pathology*, 27, 585-589 (1974)
220. Newsom, S. W. B. and Walsingham, B. M., 'Sterilization of the biological safety cabinet', *Journal of Clinical Pathology*, 27, 921-924 (1974)
221. *Health: Dust in Industry*. Technical Data Note 14, Department of Employment, HMSO, London
222. *Building Regulations 1972/73*. Guidance Note. Structural fire precautions. Department of the Environment, Welsh Office, HMSO, London (1975)
223. D.E.S. Safety Series No. 2. *Safety in Science Laboratories*. Department of Education and Science, HMSO, London (1973)

Index

Air-flow sensors, 61, 67, 80
Animal rooms, 8, 9
Automatic drench system, 44, 48

Broken glass, disposal of, 118

Change-rooms, 6
Chemicals, storage of, 63, 104
Cleaners, 35
Cleaning of floors, 16
Cleanliness, 11
Compartmentation, 34, 80

Decontamination, 12, 13
Detectors, reliability of, 43, 48
Dilution of fumes, 87
Disabled persons, 30, 34
Dispensary for solvents, 39, 110
Drains, 25, 39
Draughts, avoidance of, 73
Drench systems, automatic, 44, 48
Ducts, drainage of, 77, 78, 82

Electronic equipment, segregation of, 6
Explosions, danger of, in glove-boxes, 72
Explosives, storage of, 39, 115

Fans, choice of, 86
Fans, drainage of, 82
Filters, 73, 79, 95
Finishes, surface, 10 et seq.
Fire, gas cylinders in, 112
Fire dampers, intumescent honeycomb, 73, 81
Floods, containment of, 12
Floors, cleaning of, 16
Floors, materials for, 14, 15
Fume-cupboards, 51, 58 et seq.
Fume-cupboards, ventilation rates, 55 et seq.

Gas, supplies of, 27
Glass for fume-cupboards, 61
Glass, broken, disposal of, 118
Glove-boxes, 8, 68, 130

Hand-washing, 105
Heat, radiant, fire spread by, 85
Heating, choice of, 12
Hydrofluoric acid, 64
Hydrogen sulphide, 87

Incinerators, 39, 117

Lead shielding, 1, 8, 63, 123, 124
Lifts, fire and emergency, 30

Macerators, 117
Maintenance, 5, 35

Noise, 26, 82, 89

Oil, waste, disposal of, 118
Oxygen, excess of, 28
Oxygen, lack of, 33, 38, 48

Paints, choice of, 10
Perchloric acid, 64, 85
Plastic bag sealing equipment, 24, 79

Radiant heat, fire spread by, 85
Rainwater, 82, 87

Safety cabinets, testing of, 133
Security, 32, 35, 114
Segregation of areas, 2, 4
Shielding, lead, 1, 8, 63, 123, 124
Sinks, 22, 23
Solvent cupboards, 110
Special enclosures, ventilation of, 67 et seq.

Index

Spillage, containment of, 12
Statutory requirements, 3, 27, 38, 39, 57, 58, 106, 115, 117, 118, 124, 125, 126, 137, 138
Surfaces, work, 16 et seq.

Vehicles, access for, 31, 104, 109, 114, 119
Ventilation, calculation of rates of, 51 et seq.
Vermin, exclusion of, 12, 112, 115

Walls, materials for, 10, 11
Waste, microbiological, 133
Waste, radioactive, 116
Waste disposal unit, 24
Waste-pipes, 25
Waste solvent, disposal of, 39, 111
Weather, effect of, on fume dispersal, 88, 92

Zoning of buildings, 44